Water Informatics for Water Resource Management

The primary focus of the book series is to understand digital technologies and solutions thereby enabling them to correlate the solutions with their region-specific problems and finally apply them to the Water Resource Management. This will help to generate novel research and innovative solutions. This domain of Water Informatics is growing at a very fast rate and its potential has been realized since it harnesses the use of digital technologies to leverage the applications offered for Water Resource Management.

The book series is tailored to meet the following learning objectives:

- Learn Machine Learning Techniques for data-driven modeling for management of Water Utilities.
- Learn IOT technologies for Smart Water Management.
- Learn Artificial Neural Networks for decision making in Water Utilities.
- Learn Risk Analysis, Fuzzy Logic and management of uncertainty in Water Utilities.
- Learn Block-Chain Technology for effective allocation of Water Resources.
- Learn Cyber-Security essential techniques for Smart Water Services.
- Learning Evolutionary Methods for Water Management.

Shray Pathak · Manish Pandey
Editors

Geospatial Technology and Advanced Instrumentation in Addressing Water Resources Planning and Management

Springer

Editors
Shray Pathak
Department of Civil Engineering
Indian Institute of Technology Ropar
Rupnagar, Punjab, India

Manish Pandey
Department of Civil Engineering
Indian Institute of Technology Kharagpur
Kharagpur, West Bengal, India

ISSN 2731-9105 ISSN 2731-9113 (electronic)
Water Informatics for Water Resource Management
ISBN 978-981-92-1601-7 ISBN 978-981-92-1602-4 (eBook)
https://doi.org/10.1007/978-981-92-1602-4

This Springer imprint is published by the registered company Springer Nature Singapore Pte Ltd.
The registered company address is: 152 Beach Road, #21-01/04 Gateway East, Singapore 189721, Singapore

Preface

Water is fundamental to human survival, ecological balance, and sustainable development. Yet, growing population pressures, rapid urbanization, climate variability, and competing sectoral demands have made water resources increasingly vulnerable across the globe. Effective planning and management of water resources now require innovative approaches that integrate scientific knowledge, technological advancements, and data-driven decision-making. In this context, geospatial technology and advanced instrumentation have emerged as indispensable tools for understanding, monitoring, and managing water systems with greater precision and efficiency.

This book, *Geospatial Technology and Advanced Instrumentation in Addressing Water Resources Planning and Management*, aims to present a comprehensive overview of the modern techniques that are transforming the way water resources are assessed and governed. The chapters in this volume bring together contributions from scholars, scientists, engineers, and practitioners who offer diverse perspectives on both theoretical developments and practical applications. The book covers a wide range of topics, including watershed management, groundwater assessment, flood risk analysis, drought monitoring, water quality evaluation, irrigation planning, and climate change adaptation. Through case studies, methodological discussions, and technological insights, the contributors demonstrate how integrated approaches can improve resource efficiency, strengthen resilience, and support sustainable water governance.

A key objective of this book is to bridge the gap between research and practice. By highlighting real-world applications and emerging trends, it seeks to support decision-makers, planners, and stakeholders in adopting technologies that enhance transparency, accuracy, and long-term planning. Furthermore, the volume underscores the importance of interdisciplinary collaboration, recognizing that effective water management lies at the intersection of engineering, environmental science, geoinformatics, policy, and community engagement. This book is intended for researchers, graduate students, professionals, and policymakers engaged in hydrology, environmental management, civil engineering, geography, and related fields. The successful completion of this volume would not have been possible without the dedication of the contributing authors, whose expertise and commitment have enriched its content. We extend our sincere appreciation to the reviewers for their valuable insights and constructive feedback, and to the publishing team for their guidance throughout the production process. We also acknowledge the institutions and organizations that continue to support research and capacity building in water resources management.

As water challenges grow more complex, the integration of geospatial technology and advanced instrumentation offers a pathway toward smarter, more adaptive management strategies. We hope this book serves as a meaningful resource for advancing knowledge, fostering collaboration, and promoting sustainable solutions for present and future generations.

Rupnagar, India Dr. Shray Pathak
Kharagpur, India Dr. Manish Pandey

Contents

About the Editors

Dr. Shray Pathak is working as an Assistant Professor in the Department of Civil Engineering, Indian Institute of Technology Ropar, Rupnagar, Punjab, India. Dr. Pathak has received the Bachelor's degree in Civil Engineering from the National Institute of Technology (NIT), Kurukshetra, India in June 2012 and completed his Master's and Ph.D. degree in Civil Engineering from the Indian Institute of Technology (IIT), Roorkee, Uttarakhand, India in June 2014 and February 2019 respectively. His research interests include hydrology, remote sensing, GIS, urban sustainability and climate change. Earlier, he was employed as an Assistant Professor in the Department of Civil Engineering, Netaji Subhas University of Technology, New Delhi from Dec 2021 to Sept 2023. He has a post-doctoral experience of 2.5 years in School of Geographic Sciences, East China Normal University (ECNU), Shanghai, China (June 2019 to Dec 2021) and worked on the risk assessment analysis of urban flooding in coastal and non-coastal regions under extreme climate scenarios. Till date he has amassed over 45 publications, spanning peer-reviewed journals, conferences, and book chapters. He has received the 2022 Best Practice Oriented Paper for the Journal paper published in *ASCE Journal of Hazardous, Toxic and Radioactive Waste* awarded by the American Society of Civil Engineers (ASCE).

Dr. Manish Pandey is an Assistant Professor in the Department of Civil Engineering at the Indian Institute of Technology Kharagpur. His research interests span experimental hydraulics, sediment transport, river training works, hydraulic structures, and open-channel flow dynamics. Dr. Pandey has authored more than 60 publications, including peer-reviewed journal papers, book chapters, and conference proceedings. In recognition of his significant contributions to the field, he received the Best Reviewer Award from the Irrigation and Drainage Journal of the American Society of Civil Engineers (ASCE) in 2023, as well as the Prof. R. J. Garde Young Researcher Award in the year 2024. Dr. Pandey is an Associate Editor for the *ISH Journal of Hydraulic Engineering*, *Environmental Development and Sustainability*, and the *Journal of Applied Science and Engineering*. He is also a member of the editorial team for the *Journal of Applied Water Engineering and Research (IAHR)*. Dr. Pandey has collaborated with prominent universities and research institutions, including the University of Glasgow, the University of Bradford, Southeast University in China, and National Chung Hsing University (NCHU) in Taiwan.

Assessment of Potential Glacial Lake Outburst Flood in Uttarakhand Using Remote Sensing and GIS

Nischal Sharma and Shray Pathak

Abstract Glacial lake outburst floods are a serious hydrometeorological hazard in the rapidly changing Himalayan landscape. This study evaluates the GLOF hazard potential of Kedar Tal, a high-altitude glacier lake in the Gangotri Region of Uttarakhand, India. Multi-temporal satellite imagery (Landsat 5, Sentinel-2) and digital elevation models (SRTM) were utilized to assess lake expansion, terrain attributes, and surface water connectivity to downstream channels. Possible triggering factors were assessed datasets such as USGS seismic records and GPM rainfall were integrated. The lake's area expanded by 53.71% between 1998 and 2025, according to the results and the surrounding slopes are steeper than 35°–50°, indicating a high risk of avalanches and landslides. Although there has never been a GLOF at Kedar Tal, the environmental factors that have existed over time suggest that there is a moderate to high chance of one. The research suggests the incorporation of Kedar Tal in regional early warning and disaster mitigation plans emphasizes the need of early detection and risk assessment for lesser-studied Himalayan glacial lakes.

Keywords Glacial lake outburst flood (GLOF) · Terrain analysis · Lake expansion · Slope instability · Gangotri Region

1 Introduction

Hydrometeorological hazards includes flash floods, river floods, glacial lake outburst floods (GLOFs), dam breach floods, and draughts, influenced by weather patterns (Kumar & Murugesh Prabhu, 2012; Westen, 2013). Climate change in the Himalayan region in the recent decades has accelerated glacier retreat, leading to substantial rise in the glacial lakes (Liu et al., 2014). Glacierized mountains, including the Himalayas,

N. Sharma (✉)
Masters Student, Department of Geography, Panjab University, Chandigarh, India
e-mail: nischalsharma915@gmail.com

S. Pathak
Department of Civil Engineering, Indian Institute of Technology, Ropar, Rupnagar, Punjab, India

S. Pathak and M. Pandey (eds.), *Geospatial Technology and Advanced Instrumentation in Addressing Water Resources Planning and Management*, Water Informatics for Water Resource Management, https://doi.org/10.1007/978-981-92-1602-4_1

are more sensitive to climate change compared to the flood warm plains (Khadka et al., 2023). 65% of the ice mass of High Mountain Asia (HMA) is predicted to dissipate by the end of this century citing climate model projections driven by high emission scenarios (Kraaijenbrink et al., 2017).

Glacial lake outburst flood (GLOF) is the rapid release of a large amount of water retained in a glacial lake. Glacial lakes are formed by the thinning and retreat of glaciers which are dammed by glacial deposits (Sattar et al., 2019). During glacier retreat, overdeepenings from prior glacier beds have been exposed in numerous places, forcing meltwater accumulate as glacial lakes (Taylor et al., 2023). These moraines, some of which contain a melting ice core, are built from rock debris transported by glaciers. When they fail, large volumes of stored water can be released, producing glacial lake outburst floods (GLOFs) (Harrison et al., 2018; Veh et al., 2018). GLOFs have extraordinary erosion and transport capacity and severe peak discharges, a capacity frequently several times greater than the maximum discharges of floods caused by hydrometeorological factors (Lützow et al., 2023). GLOF consists of mixture of water and sediment which are capable of travelling tens of kilometers to more than 100 km at velocities exceeding tens of kilometers per hour and cause serious threat with their sudden outburst due to their high magnitude discharge and long runout distance (Worni et al., 2014). Glacial lake outburst floods can have catastrophic downstream effects on water systems, infrastructure, and human populations (Maskey et al., 2020). For instance, Kedarnath disaster of Uttarakhand, India (2013) that occurred due to breaching of Chorabari Lake (Champati Ray et al., 2016) and floods from DIG Tsho (1985) (Gurung et al., 2017). Advanced techniques that can handle complexity of terrain, hydrology and climate fluctuations are necessary for assessment, monitoring, and management of such high-risk glacier ecosystems. GIS-based spatial analysis makes it feasible to accurately monitor the expansion of lake, catchment characteristics, and flood susceptibility, even in inaccessible high-altitude regions (Ahmed et al., 2021).

This chapter focuses on assessment of potential glacial lake outburst flood in Uttarakhand's Gangotri Region's, Kedar Lake using Remote Sensing and GIS. The study examines the changes in lake area, slope characteristics and potential triggers that can cause downstream impact. Most of the researches have been done on larger lakes which have a clear expansion histories. There are no extensive researches on Kedar Tal amid the expanding literature on Himalayan glacial lakes. Such lakes are potentially vulnerable and hazardous to the settlements downstream and infrastructure.

2 Study Area

Kedar Tal is situated in Uttarakhand, Garhwal Himalayas at an elevation of 4729 m above the mean sea level at 30.8156° N latitude and 78.9311° E longitude. The lake is situated in the group of Gangotri peaks, the Himalayas surrounding the Gangotri town of Uttarkashi district of Uttarakhand State. It is fed by snowfall of surrounding

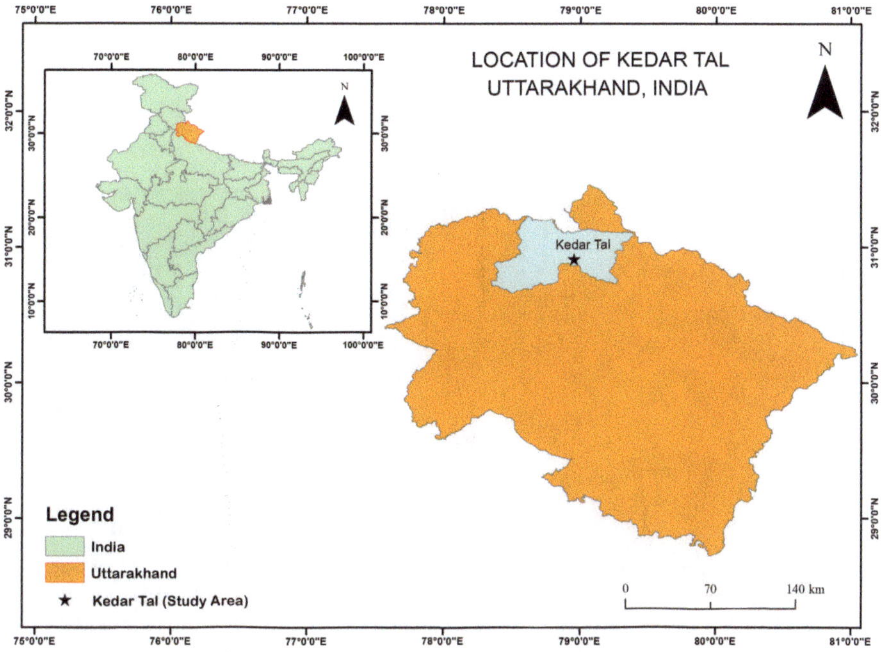

Fig. 1 Map of India

peaks. It is one of the popular trekking destinations, located 17 km from the Gangotri. It lies in the upper catchment area of Bhagirathi Basin. The lake's environment is unstable and susceptible to rapid changes in geomorphology. The lake and the glacier surrounding it are part of Gangotri glacier system which considerably enhances the Ganga River's perennial flow (Figs. 1 and 2).

2.1 Datasets

2.1.1 Satellite Imagery

Datasets used in this study were optical satellite imagery of Landsat 5 thematic mapper level 1 with resolution of 30 m for the year 1998; Sentinel-2 level 2 optical imagery resolution of 10 m for the year 2025. Landsat Data was downloaded from United States Geological Survey (https://earthexplorer.usgs.gov) and Sentinel Data was downloaded from Copernicus Open Access Hub (https://scihub.copernicus.eu). Using false color composites,

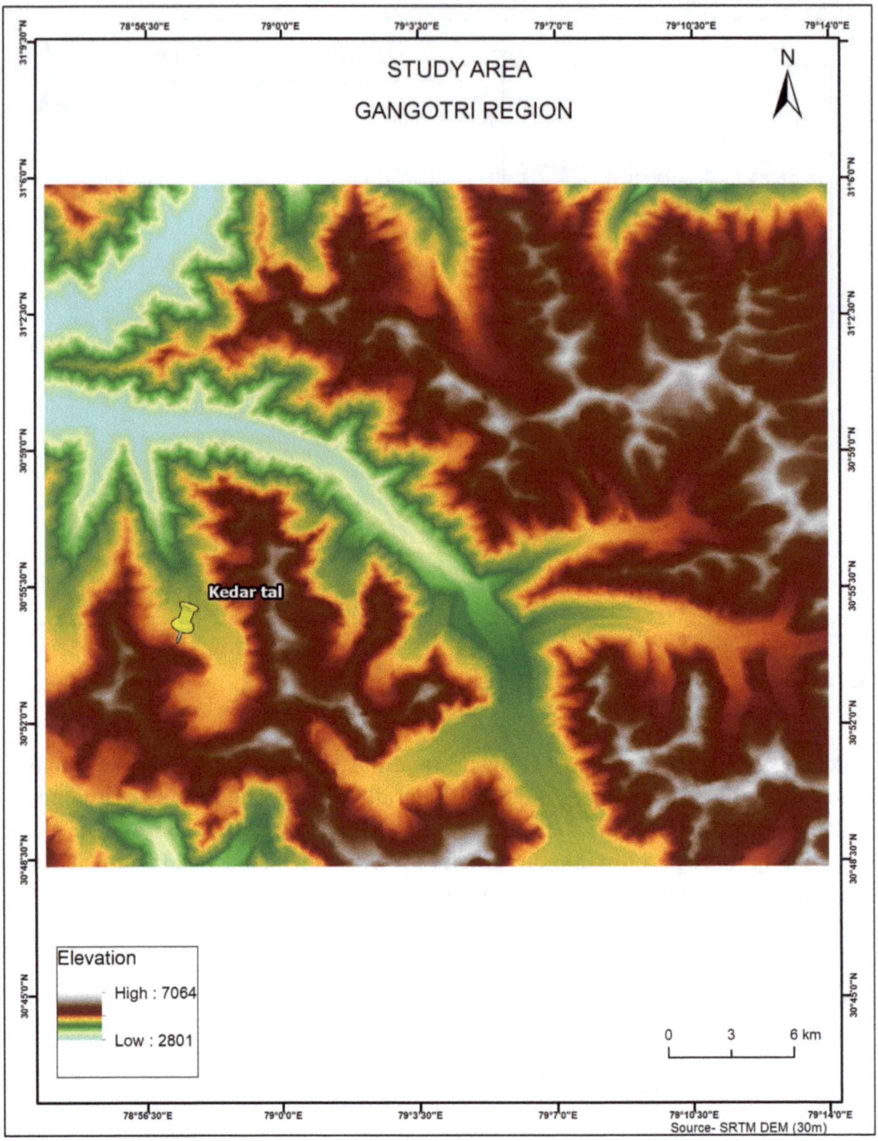

Fig. 2 Gangotri region

2.1.2 Digital Elevation Model

Shuttle Radar Topography Mission (SRTM) DEM was used in the study with a spatial resolution of 30 m. It was acquired from NASA Open Topography (https://opentopography.org) determination of slope, elevation for terrain evaluation, potential downstream flood assessment was done using SRTM DEM.

2.1.3 Rainfall Data

A time-averaged rainfall data was downloaded from NASA-GIOVANNI portal. It represents surface precipitation from time range of June 1998 until June 2025. Highest quality Global Precipitation Method GPM product was used in this study. The Giovanni rainfall data has resolution of 10 km.

2.1.4 Earthquake Data

United States Geological Survey (USGS) earthquake catalogue was used for the extracting earthquake data around the Kedar Tal region. The earthquakes of magnitude greater than 2.5 were taken into account from time period of 1998–2025.

Arc GIS 10.8 software was used for the analysis. All the datasets were projected to WGS 84 and UTM Zone 44N for spatial consistency.

3 Methodology

3.1 Preparation and Visualization of Satellite Images

Landsat 5 TM (1998) and Sentinel-2 MSI (2025) were preprocessed. Band Composition and image enhancement was done. For Landsat 5, bands 5 (SWIR), band 4 (NIR), and band 3 (Red) were combined to generate a false color composite to identify the glacial lake. For Sentinel-2, band 8 (NIR), band 4 (Red), and band 3 (Green) were combined to generate a false color composite.

3.2 Lake Boundary Delineation

The Kedar Tal was delineated manually for both the satellite images of Landsat and Sentinel. The digitization was performed using Arc GIS on FCC images. The area of both the years was calculated along with the rate of change for the temporal analysis of the lake through the years 1998 and 2025.

3.3 Terrain Characterization Using SRTM DEM

SRTM DEM with spatial resolution of 30 m resolution was used to analyze the slope of the area and assess the potential triggers that can occur and contribute to the glacial lake outburst flood. In order to analyze the possible flood routes and flow accumulation downstream D8 method was used.

3.4 *Rainfall Intensity Evaluation Using GPM Data*

GPM data through the years 1998 and 2025 was downloaded from GIOVANNI for the clipped area surrounding Gangotri and its peaks was analyzed. The raster data was then reprojected to UTM Zone 43. It revealed the average monsoonal precipitation rates in the Gangotri and Kedar Tal region. This contributed in the analysis of rainfall as potential trigger.

3.5 *Earthquake Data Interpretation*

Seismic Data sourced from USGS Earthquake Catalogue ranging the events of magnitude > 2.5 for the clipped area of Gangotri and its surrounding peaks including Kedar Tal was downloaded in KML and mapped in ArcGIS. Earthquake epicenters were visualized to assess their proximity to the glacial lake and potential influence on moraine stability.

3.6 *Spatial Integration and GLOF Hazard Assessment*

Qualitative interpretation was conducted to evaluate the hazard potential of the glacial lake Kedar Tal. All spatial layers including the lake boundaries including digitized lake boundaries, slope maps, flow paths, stream order, rainfall intensity, and earthquake epicenters based on expansion of the lake area, slope steepness, downstream impact, and presence of external triggers.

4 Results and Discussion

4.1 *Temporal Variation in Lake Area*

There was noticeable increase observed in the digitization of Kedar Tal region. In 1998 the area of the lake was 74,570.66 m^2 whereas in the year 2025 the lake area observed was 114,625.87 m^2. The rate of change of the lake area was calculated using the formula

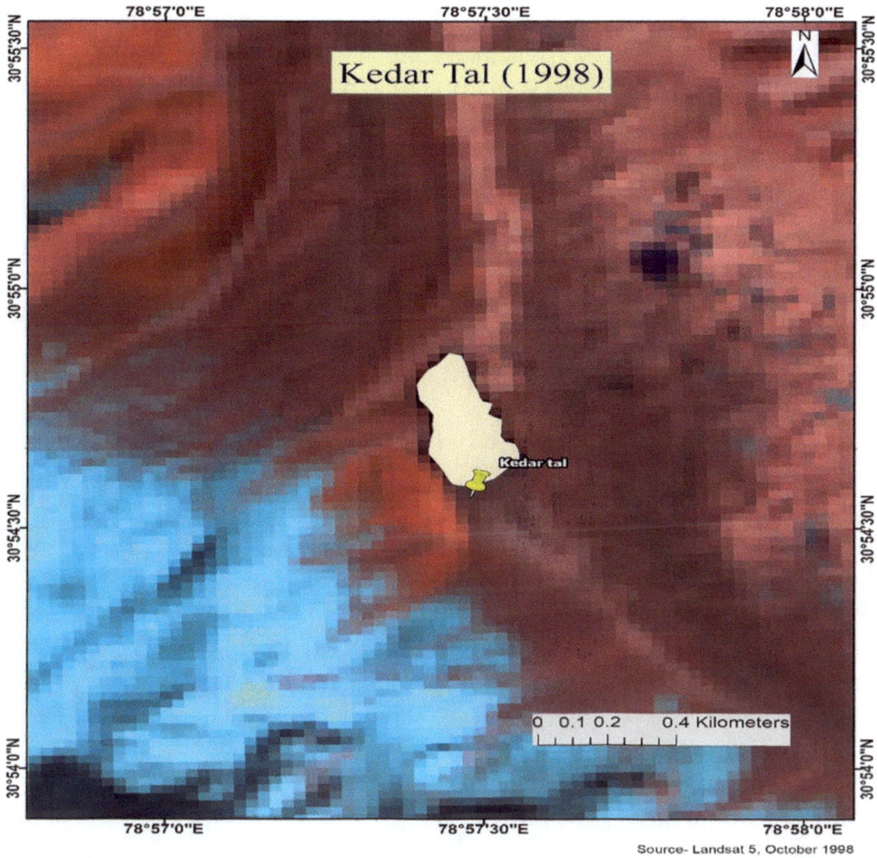

Fig. 3 Kedar Tal in 1998

$$\text{Rate of change} = (\text{new area} - \text{old area}/\text{new area}) * 100$$

The rate of change of the lake area is 53.71%.

This shows temporal and spatial variation of the glacial lake. It has increased double in size through the timespan of 27 years. The expansion pattern is similar to other glacial lakes in the Central Himalaya, where glacier retreat and meltwater accumulation have been widely reported (Worni et al., 2013) (Figs. 3 and 4).

4.2 Extraction of Terrain Attributes from DEM

The slope of the area surrounding the Gangotri Region and Kedar Tal was generated. The region is characterized by steep slopes around the Kedar Tal particularly in South

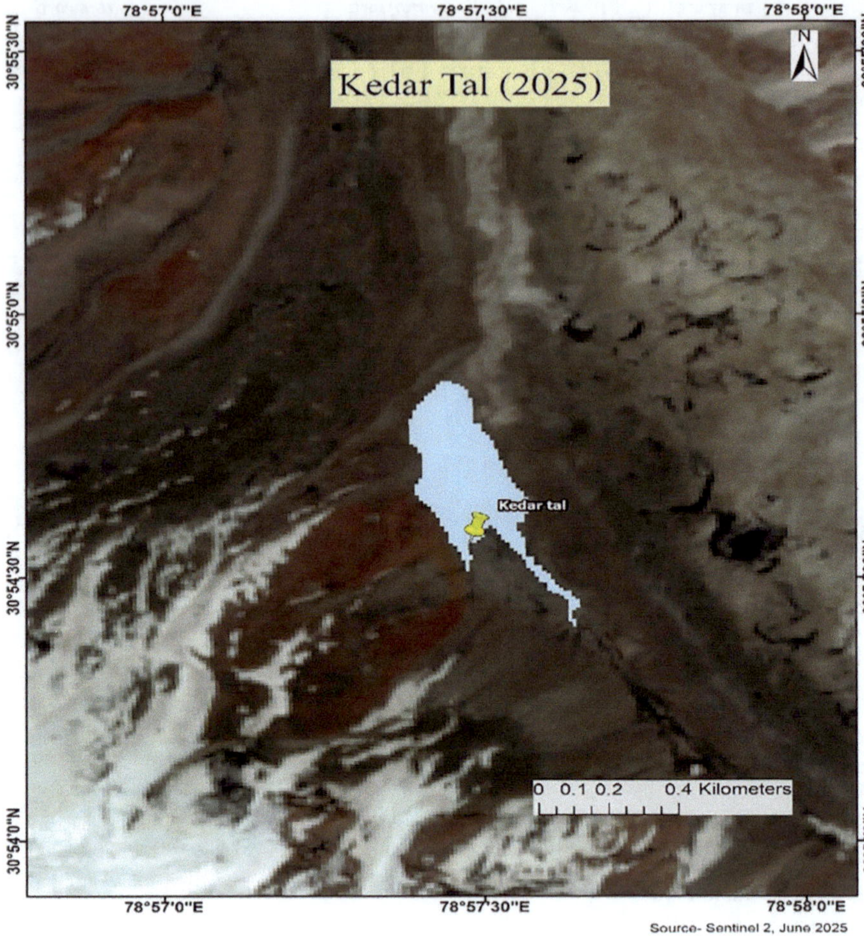

Fig. 4 Kedar Tal in 2025

and South East. The area with slope of 30° surrounding the Kedar Tal is susceptible to landslides and the slope of more than 40° is susceptible to avalanches shown in yellow to orange. The steep slope makes the lake vulnerable to debris flow, rockfall which can directly cause the breach of the lake, weakening of unstable moraine dam and eventually the GLOF.

4.3 Stream Network and Hydrological Connectivity

Using the flow accumulation layer, the Kedar Tal region's stream network has been extracted out of the SRTM DEM. The outlet of Kedar Tal feeds directly into a first-order stream. This indicates that a potential glacial lake outburst flood (GLOF) would rapidly enter a higher-order drainage system, causing fast and wide downstream propagation. Kedar Tal is a high-altitude lake, its stream connectivity increases the hazard potential to downstream valleys and settlements. The stream order turns out crucial for GLOF risk assessment and early warning system planing (Figs. 5, 6 and 7).

4.4 Rainfall Intensity Trends

Time-averaged GPM IMERG (1998–2025) precipitation data was obtained from NASA Giovanni and interpolated using ArcGIS's Inverse Distance Weighting (IDW) methodology in order to display long-term rainfall variance. The resulting map classifies rainfall intensity into five ranges. Kedar Tal lies in the orange rainfall zone indicating moderately high precipitation. The possible impact of prolonged rainfall on lake volume and moraine stability is shown by this spatial pattern which is crucial for determining the danger of GLOF (Fig. 8).

4.5 Earthquake

Moderate earthquakes were recorded around the region of Kedar Tal between 1998 and 2025. While no known GLOF event has occurred right after the earthquakes in the region, but earthquakes can destabilize the moraines. Kedar Tal lies in the low to moderate. Earthquake is not a triggering factor for a potential Glof (Fig. 9).

4.6 Hazard Interpretation

The combination of lake expansion, steep terrain, intense monsoonal rainfall, and regional seismicity suggests that Kedar Tal presents a moderate to high GLOF hazard potential. According to the findings, the lake is situated in a glacial basin at a high elevation, with steep to extremely steep slopes (35°–50°), particularly on its eastern and southern sides.

The rainfall analysis, based on GPM IMERG time-averaged data (1998–2025), revealed that Kedar Tal lies within a moderately high precipitation zone. While the rainfall alone is not sufficient to trigger a GLOF, it can contribute to moraine

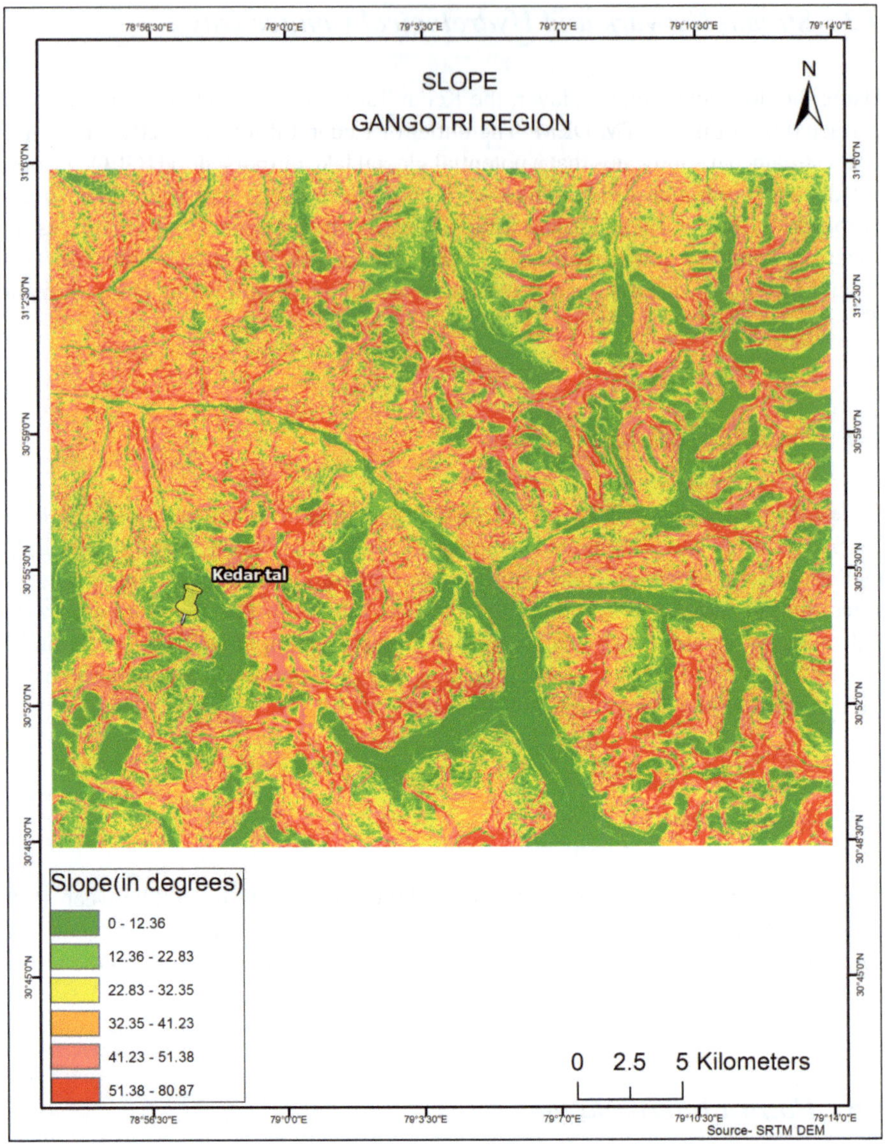

Fig. 5 Slope around Kedar Tal region

saturation, glacial melt, and long-term lake volume increase especially during the monsoon season. The stream network and flow direction analysis, derived using the D8 algorithm, show that the lake drains into a first-order stream which quickly connects to higher-order channels downstream. This indicates that any sudden release of water would be rapidly transmitted to lower elevations, increasing the flood reach and intensity. Furthermore, the presence of seismic activity in the region, as shown

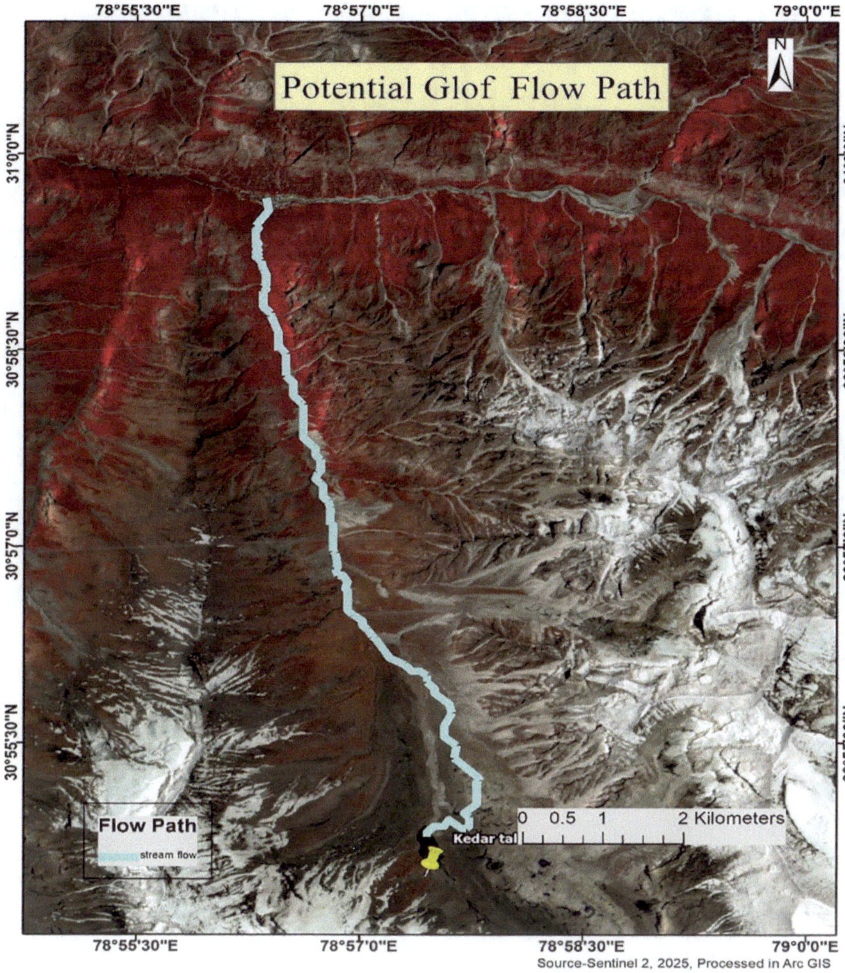

Fig. 6 Stream network

by historical earthquake data from USGS, adds an additional dimension to the hazard scenario. A moderate earthquake in proximity to the lake could act as a triggering factor, potentially causing moraine destabilization or inducing slope failure into the lake basin. As such, Kedar Tal warrants continued monitoring, possible field-based moraine stability assessments, and its inclusion in regional GLOF early warning and disaster planning frameworks Kedar Tal has a number of hydrometeorological and geophysical features that, although not particularly severe on their own, add up to a multi-hazardous environment.

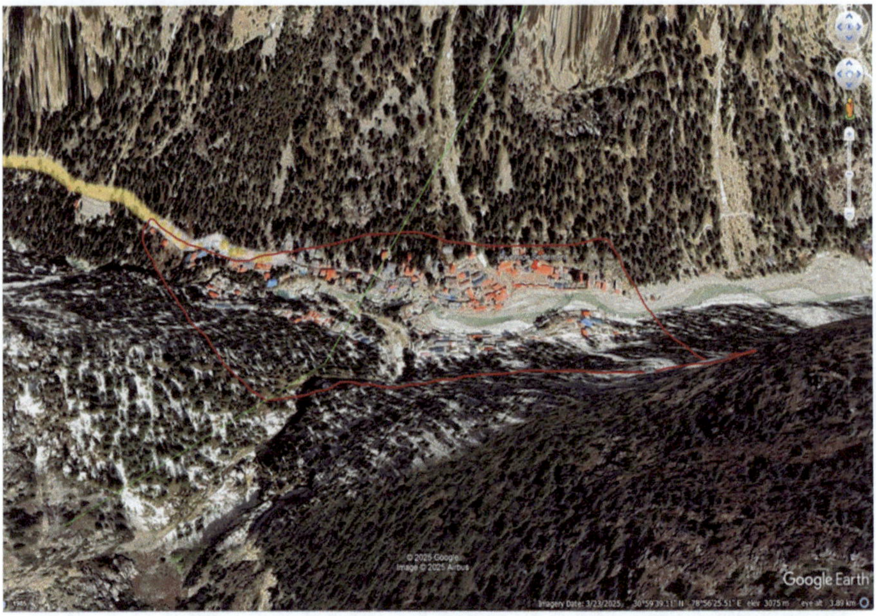

Fig. 7 Downstream settlements

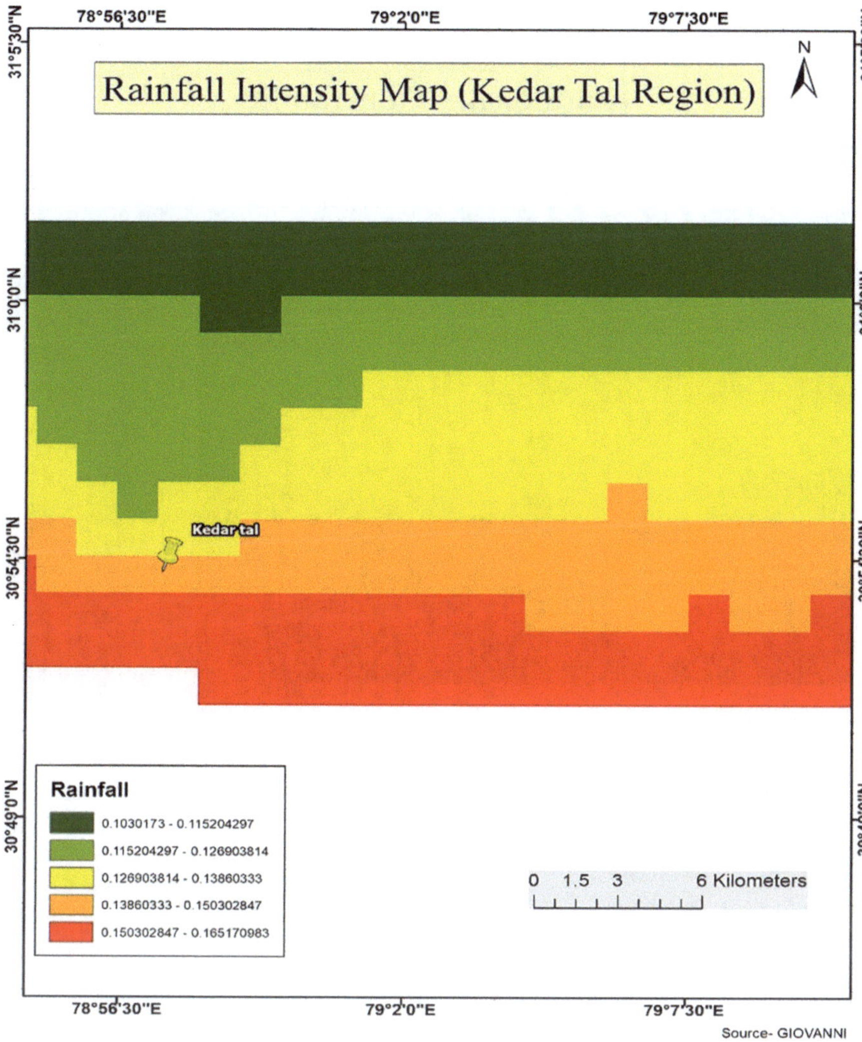

Fig. 8 Rainfall intensity

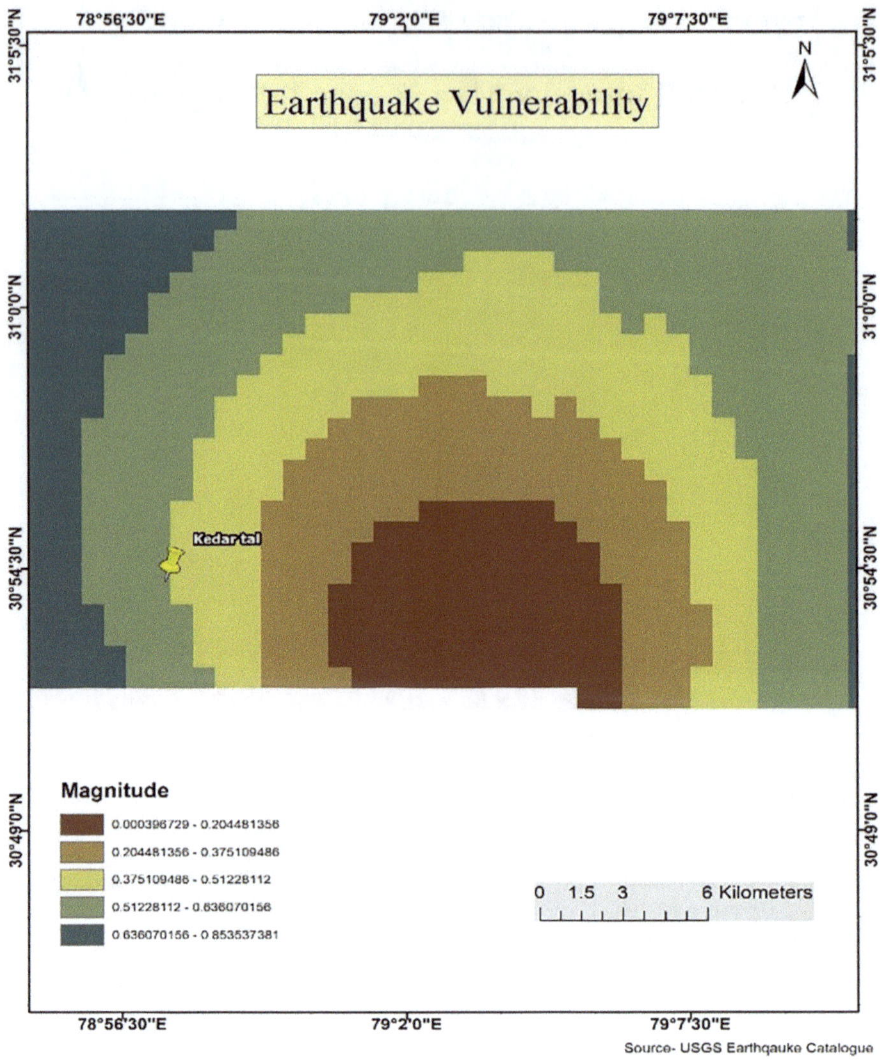

Fig. 9 Earthquake vulnerability

5 Conclusion

This study assessed the possible glacial lake outburst flood (GLOF) risk related to Kedar Tal, a high-altitude glacial lake in Uttarakhand, India's Garhwal Himalaya. Several risk-inducing elements were evaluated using an integrated geospatial method that included Hydro-spatial analysis, remote sensing, and DEM-based terrain analysis. The findings revealed that Kedar Tal is positioned within a steep mountainous basin, with surrounding slopes (particularly to the south and southeast) surpassing

35°–50°. The loose glacier and moraine debris that makes up the majority of these slopes is prone to breakdown. Slope instability, mapped from the SRTM DEM, emerges as a key potential GLOF trigger. Rainfall data from GPM IMERG (1998–2025) showed that the area receives moderately high average precipitation, especially during monsoon months.

As climate change continues to reshape Himalayan glacier dynamics, the potential for glacial lake outburst floods (GLOFs) is expected to increase in both frequency and severity. Studies have shown that warming temperatures contribute to accelerated glacial melt, rapid expansion of glacial lakes, and increased hydrometeorological instability in the region (King et al., 2019). For an assessment of the change in Glacial Lake's nature, aerial extent and repeated monitoring is necessary (Jain et al., 2025). Raising locals' awareness is just as crucial to deterring infrastructure development and real estate investment for conservation and promotion of tourism in such areas (Bajracharya et al., 2007).

Although Kedar Tal currently shows no immediate signs of instability, its topographical and hydrological settings place it in a potentially vulnerable category. Therefore, future research and disaster risk management strategies should emphasize long-term monitoring, lake volume estimation, and early warning infrastructure.

Incorporating high-resolution DEMs, multi-temporal satellite imagery, and ground-based hydrometeorological stations can significantly improve hazard prediction models (Veh et al., 2019). Integrating community-level disaster preparedness, particularly for trekking routes and downstream villages, is critical for improving regional resilience (Haeberli et al., 2016).

References

Ahmed, R., Wani, G. F., Ahmad, S. T., Sahana, M., Singh, H., & Ahmed, P. (2021). A review of glacial lake expansion and associated glacial lake outburst floods in the Himalayan region. *Earth Systems and Environment, 5*(3), 695–708.

Bajracharya, B., Shrestha, A. B., & Rajbhandari, L. (2007). Glacial lake outburst floods in the Sagarmatha region. *Mountain Research and Development, 27*(4), 336–344.

Champati Ray, P. K., Chattoraj, S. L., Bisht, M. P. S., Kannaujiya, S., Pandey, K., & Goswami, A. (2016). Kedarnath disaster 2013: Causes and consequences using remote sensing inputs. *Natural Hazards, 81*(1), 227–243.

Gurung, D. R., Khanal, N. R., Bajracharya, S. R., Tsering, K., Joshi, S., Tshering, P., et al. (2017). Lemthang Tsho glacial Lake outburst flood (GLOF) in Bhutan: Cause and impact. *Geoenvironmental Disasters, 4*(1), 17.

Haeberli, W., Buetler, M., Huggel, C., Friedli, T. L., Schaub, Y., & Schleiss, A. J. (2016). New lakes in deglaciating high-mountain regions–opportunities and risks. *Climatic Change, 139*(2), 201–214.

Harrison, S., Kargel, J. S., Huggel, C., Reynolds, J., Shugar, D. H., Betts, R. A., et al. (2018). Climate change and the global pattern of moraine-dammed glacial lake outburst floods. *The Cryosphere, 12*(4), 1195–1209.

Jain, S. K., Jain, S. K., & Ahmad, R. (2025). Glacial lake outburst floods (GLOFs): Causes, modeling, and mitigation in Indian Himalayan region. *GeoJournal, 90*(2), 85.

Khadka, N., Chen, X., Sharma, S., & Shrestha, B. (2023). Climate change and its impacts on glaciers and glacial lakes in Nepal Himalayas. *Regional Environmental Change, 23*(4), 143.

King, O., Bhattacharya, A., Bhambri, R., & Bolch, T. (2019). Glacial lakes exacerbate Himalayan glacier mass loss. *Scientific Reports, 9*(1), 18145.

Kraaijenbrink, P. D., Bierkens, M. F., Lutz, A. F., & Immerzeel, W. W. (2017). Impact of a global temperature rise of 1.5 degrees Celsius on Asia's glaciers. *Nature, 549*(7671), 257–260.

Kumar, B., & Murugesh Prabhu, T. S. (2012). Impacts of climate change: Glacial lake outburst floods (GLOFs). In *Climate change in sikkim patterns, impacts and initiatives*. Information and Public Relations Department, Government of Sikkim, Gangtok.)

Liu, J. J., Cheng, Z. L., & Su, P. C. (2014). The relationship between air temperature fluctuation and Glacial Lake Outburst Floods in Tibet, China. *Quaternary International, 321*, 78–87.

Lützow, N., Veh, G., & Korup, O. (2023). A global database of historic glacier lake outburst floods. *Earth System Science Data Discussions, 2023*, 1–27.

Maskey, S., Kayastha, R. B., & Kayastha, R. (2020). Glacial lakes outburst floods (GLOFs) modelling of Thulagi and lower Barun glacial lakes of Nepalese Himalaya. *Progress in Disaster Science, 7*, 100106.

Sattar, A., Goswami, A., & Kulkarni, A. V. (2019). Hydrodynamic moraine-breach modeling and outburst flood routing—A hazard assessment of the South Lhonak lake, Sikkim. *Science of the Total Environment, 668*, 362–378. https://doi.org/10.1016/j.scitotenv.2019

Taylor, C., Robinson, T. R., Dunning, S., Rachel Carr, J., & Westoby, M. (2023). Glacial lake outburst floods threaten millions globally. *Nature Communications, 14*(1), 487.

Veh, G., Korup, O., Roessner, S., & Walz, A. (2018). Detecting Himalayan glacial lake outburst floods from Landsat time series. *Remote Sensing of Environment, 207*, 84–97.

Veh, G., Korup, O., von Specht, S., Roessner, S., & Walz, A. (2019). Un-changed frequency of moraine-dammed glacial lake outburst floods in the Himalaya. *Nature Climate Change, 9*(5), 379–383.

Van Westen, C. J. (2013). Remote sensing and GIS for natural hazards assessment and disaster risk management

Worni, R., Huggel, C., Clague, J. J., Schaub, Y., & Stoffel, M. (2014). Coupling glacial lake impact, dam breach, and flood processes: A modeling perspective. *Geomorphology, 224*, 161–176.

Worni, R., Huggel, C., & Stoffel, M. (2013). Glacial lakes in the Indian Himalayas—From an area-wide glacial lake inventory to on-site and modeling based risk assessment of critical glacial lakes. *Science of the Total Environment, 468*, S71–S84.

Emotional Labour Under Smart Water Systems: Job Satisfaction and Burnout Implications

Diksha Negi and Manisha Semwal

Abstract This chapter explores increased emotional labour required by frontline workers of water utilities who work in technologically advanced settings. Since smart water systems (SWS) combine Supervisory Control and Data Acquisition (SCADA), internet of things (IoT) and real-time analytics in water governance, the psychological well-being of affected employees who administer the systems is being highly compromised. Case studies to help us examine the effectiveness of cognitive overload, emotional regulation, and technological surveillance, on job satisfaction and burnout as conducted in Spain, India, Singapore and the USA. Based on emotional labour theory and the Job Demands-Resources (JD-R) model, the chapter contains practical recommendations that can create emotional resilience and sustainable workforce well-being in smart water utilities.

Keywords Burnout · Algorithmic management · Emotional labour · Job satisfaction · Job demands resources model · Supervisory Control and Data Acquisition (SCADA) smart water systems · Frontline utility workers

1 Introduction

Climate change, scarcity of resources, urbanization, and quest to attain a higher degree of operational efficiency have triggered a fast-paced digitalization of urban water systems. The embrace of smart water systems (SWS) has emerged as a key measure dealing with the global cities with the aim of achieving water security, efficient use of resources, as well as enhancing sources of responsiveness to infrastructure issues. It includes the real-time monitoring and control in the flow, quality and

D. Negi (✉)
Department of Management Studies, Graphic Era (Deemed to be University), Dehradun, Uttarakhand, India
e-mail: dikshanegi.phd@geu.ac.in

M. Semwal
Graphic Era (Deemed to be University), Dehradun, Uttarakhand, India

S. Pathak and M. Pandey (eds.), *Geospatial Technology and Advanced Instrumentation in Addressing Water Resources Planning and Management*, Water Informatics for Water Resource Management, https://doi.org/10.1007/978-981-92-1602-4_2

leakage, water is provided with these systems in which the high degree of deployment of Supervisory Control and Data Acquisition (SCADA), Internet of Things (IoT) devices, Geographic Information Systems (GIS) and machine learning analysis are featured.

Nevertheless, there is not a human-free price to the usage of such technologies. Front line workers in water utilities have normally worked in low intensity, mostly deterministic environments that are being transformed into data-intensive and high-intensity settings. Their jobs demand constant monitoring of dashboards, real-time alert and predictive maintenance services and they are at times monitored by surveillance systems. Although these changes make it more technically efficient, they increase emotional demands required by the workers.

In this context of this technology, the concept of emotional labour is introduced in this chapter. Emotional labour which was originally derived by Hochschild (1983) entails the control and manipulation of emotional products in accordance with the instilled expectations of an organization or a role. Operators working in water utility settings are growingly expected to keep their temper, portray compassion when handling customers, and make major decisions when under pressure, which they also must perform on a convoluted digital system.

In the context of SWS, emotional labour is no longer confined to face-to-face customer interactions. Instead, it increasingly includes managing the emotional toll of algorithmic control, constant system monitoring, digital alerts, and technostress (Tarafdar et al., 2007). Employees must suppress frustration, fatigue, and anxiety while maintaining composure in high-pressure environments shaped by machine logic and digital expectations.

As frontline workers are expected to deliver uninterrupted services and maintain system integrity, the **invisible emotional burdens** associated with their roles remain under-researched and under-recognized. These hidden stressors are compounded in environments where workers lack adequate support systems, training, or control over digital processes. Consequently, emotional exhaustion, job dissatisfaction, and burnout are becoming critical risks in smart water workplaces (Bakker & Demerouti, 2007; Maslachet al., 2001).

This chapter will discuss the nature of emotional labour in smart water settings to learn how to manage emotional strain in a job and the associated job satisfaction and burnout. The final objective of the chapter will be to present evidence-based suggestions on managing emotional strain. The results are in line with the qualitative synthesis of secondary data, where the case studies in Spain, India, and the United States have been provided. The central research question is as follows: How is emotional labour experienced in the working environment of frontline water utility employees in smart system regimes and what does it mean to their work-related well-being?

2 Emotional Labour: What It Is and Why It Matters

Emotional labour is the means of controlling the emotions to meet emotional demand of the work (Hochschild, 1983, 2003). In the water system it is the ability to act reasonably in a situation of high pressure such as a pipe breaking or contamination warning, and usually a direct or indirect witness.

Grandey (2000) and Morris and Feldman (1996) as well as researchers have expanded the construct further by including surface acting, deep acting and genuine emotional display. Decisive to the effect is that surface acting reigns pre-eminent in places, where rules and commands are scrutinized considerably e.g. in smart utilities where employees conceal stress or exhaustion to alleviate themselves with the institution.

The experience of emotional labour in public services is not a novel area of study (Guy et al., 2008) and the way it is co-constituted with emerging technologies is not fully studied. JD-R (Demerouti & Bakker, 2011) is a helpful conceptualization to discuss what happens when emotional demands (customer interactions, alert fatigue) outweigh the available resources (training, peer support) and cause a burnout.

2.1 Public Utilities Emotional Labour

Historically, emotional labour has been studied in those occupations that require interaction with customers, namely, flight attendants, service employees, health professionals. Such roles demand that employees tone down expression of their emotions to conform to institutional or even profession-oriented rules which, in most cases, create emotional dissonance and burnout.

Emotional labour has been identified in other more technical and mixed professions such as the ones in the public utilities within past few years. Employees in the water utilities tend to be dual position: not only do they have to perform technical tasks (repair pipes, coordinate distribution networks), but they also must deal with people, resolve complaints, and control situations of emergency. Such interactions call into play a high level of emotional control especially at a time when the customers are in stress or angry because there is a problem in delivering services.

In addition, the organizational culture alone has its own emotional requirement. Employees should not lose their head in times of emergencies like rupture of a main or contamination incident. There are also organizational pressures where they are obliged to exhibit commitment and compliance even when faced with inefficiency of the system or pressure by the management. In this case, emotional labour includes not only the interaction with clients but also goers through institutional expectations, relations with peers, and technology.

2.2 Cognitive Water Systems and Load

Smart water systems present a breaking point in the meaning of utility management. High-tech systems are beneficial in that they monitor the quality of the water, detecting its leaks and distributing it with limited human impact. Jesses thereof contain.

Centralized SCADA systems that encompass control overflow and distribution of water.

- IoT sensors to sense changes of pressure and flow rate and possible leakages.
- Real-time system component mapping GIS.
- Predictive maintenance, System optimization: machine learning algorithms.

Although such systems alleviate the physical work affecting the operators, they extremely heighten their mental workload. Operators must read through big amounts of data, quickly react to system warning signs, and to find solutions to the anomalies. With this digitisation of oversight, new stress engulfs the shift in the reduction of manual labour. According to the findings supplied by Steffelbauer et al. (2021), when faced with a constant flow of system alerts, a person experiences what is known as alert fatigue, or a state of mental fatigue, callousness to alerts, and heightened rates of error.

More than that, there is usually algorithmic control characterizing smart systems, according to which the efficiency of workers is measured in digital parameters. This may cause performance pressure, loss of autonomy and alienation, and all these bring about emotional stress.

2.3 Job Satisfaction and Burnout in the Workplace Amongst Utility Sectors

The growing emotional needs of the technologically facilitated utilities have direct consequences of job satisfaction and employee burnout. It seems that in the water segment, job satisfaction is not only a set of intrinsic motivators (e.g. sense of being of service to the population, control over work) but also some extrinsic variables (e.g. digital surveillance, performance measuring, work allocation).

Among the most frequently employed models of burnout evaluation are the Maslach Burnout Inventory (MBI) model that singles out three dimensions.

- Emotional exhaustion—using up emotion.
- Depersonalization and cynical beliefs about work.
- Decreased feeling of personal accomplishment–dissatisfaction and feelings of being ineffective (Maslach et al., 2001).

The MBI has been confirmed among various research conducted in the sphere of public sectors as a good indicator of long-term job disengagement and turnover motivation, particularly those involving high emotional labour (Schaufeli & Taris, 2013). With water utilities this effect is compounded by new digital monitoring and system requirements, creating an endless stressful chain in which front-line workers are perpetually in an emotional state of always-on, with little appreciation of their emotional value.

2.4 JD-R Scenario of Burnout

Job Demands-Resources (JD-R) model offers a handy framework to realize the psychology impact of being smart water environment. Job demands were defined as the physical, psychological or emotion efforts that a job demands, according to JD-R model (Demerouti et al., 2001). Where such demands are not accompanied by sufficient resources (support, training or autonomy) to compensate workers may experience burnout.

The model developed by Bakker and Demerouti (2007), known as the Job Demands-Resources (JD-R) model, can be used to view how organizational environments contribute to the incidences of burnout. This model assumes that when there is a great deal of job requirements (such as cognitive load, emotional labour, technostress) and when job resources are low (such as training, support or autonomy) the outcome is burnout and disengagement.

The requirements of the job in the perspective of smart water systems are:

- Constant alert tracking.
- Regulation of emotions in external and internal relations.
- Speedy response requirement.

Algorithmic monitoring and performance monitoring

On the contrary, resources are limited. Employees complain of poor emotional support, lack of appreciation, training on only digital tools and are rarely consulted on decision-making. The result of this imbalance is emotional exhaustion, depersonalization, and decreased personal accomplishment which are characteristics of burnout.

3 Methodology

In this chapter, the qualitative multiple-case study methodology is used. We also analysed documents and government reports, white papers on the industry and media accounts of the SWS deployments. The choice of case studies was made to include a variety of geographic and technological focal points of Europe, North America, and Asia. The emotion strain markers, the outcome of job satisfaction, and the reaction of the organization were coded by use of thematic analysis. Information can be found as follows.

- Debatable scholastic articles.
- Municipal utility reports and NGO publications, i.e. grey literature.
- Trade and industry magazine interviews (e.g. Water and Wastewater Asia).

The theme coding of the documents was also applied in the methodology to determine patterns linked to emotional labour, burn out, job satisfaction, and technology influence on the daily work routine. The cases were selected based on geographical dispersion, the maturity of technology and publicly accessible source of data.

Besides the analysis of documents, we also used theoretical frames of emotional labour theory (Hochschild, 1983) and JD-R model (Demerouti et al., 2001) to make findings more understandable.

Case Study Opinion

Barcelona, Spain Aigues de Barcelona

The water utility Barcelona has experienced success in the use of smart metering and real-time infrastructure. More than 500,000 smart meters have been connected to a central SCADA platform through which usage is continuously monitored, leakage is identified, and forecasted maintenance undertaken. Such a digital infrastructure has reduced the time to response to a leak and billing accuracy drastically.

Nonetheless, according to internal reports and the interview with the staff, things are more complicated. There is an extremely excessive amount of emotional stress that operators experience because of monitoring and system alert. They are evaluated increasingly using dashboards, and automated feedback loops, which generates the sense of being micromanaged. These employees also feel lonely since most of their interaction with the system is distant and automated, and they do not have many opportunities to interact and receive support.

Philadelphia, United States of America

To build a strong smart water ecosystem, Philadelphia Water Department has seamlessly combined SCADA technology with mobile applications and GIS. The mobile applications can be used to send issues illustrations by the customers which are sent directly to the dashboards of the operators. Although such transparency has enhanced the participation of citizens, it has increased the cognitive and emotional burden of control room personnel.

A 2021 survey conducted by the Public Utility Workers Association (PUWA) showed that employees feel increased anxiety as their actions are both fast-paced and noticeable. The operators need to follow up on incoming requests of a citizen, alerts, and internally monitored success. This flexible working with monitoring is associated with stressful nature not least because of breakdowns in the system.

Singapore—Public Utilities Board (PUB)

The Public Utilities Board (PUB) of Singapore is known to be one of the first fully integrated smart grids in the world which use robotics to inspect pipeline, IoT based metering and real time surveillance systems. Water Quality Monitoring System (WQMS), Drainage Monitoring System (DMS) and Yielding Degenerative Analysis of Assets (PAHA) are all controlled by central SCADA systems.

Although Singapore is recognized as functionally rich in technical infrastructure, its workers in PUB have been complaining of emotional burnout because of the level of expected performance and organizational culture. The technicians in the field will be in environments that are saturated with digital information with little or no human interaction thus emotional isolation. It is very procedural, as it follows the digital rules, not the human judgment, and it is psychologically taxing and emotionally anaemic.

According to the interview with PUB staff mentioned in the Smart Nation Insights (2022), employees were worried about the acceleration of the decision-making process and the necessity to bring their digital expertise up to date continuously. Training programs tend to be long but focus more on technical eloquence rather than emotional strength.

I talk to the system more than my team does. We are paper efficient, not connected,—SCADA controller, PUB, 2022.

Dehradun and Chennai, India Dehradun and Chennai, India—SCADA in Developing Contexts

SCADA and IoT have been implemented in Dehradun, and Chennai, to deal with such problems as non-revenue water, bad distribution, and outdated infrastructure. Although the technological enhancements have increased productivity, employees have not handled emotional and functional stresses of Digital transformation.

Through interviews and news reports, issues such as the aspect of alert fatigue, interference when one must answer calls in the middle of the night, and the inability to meet unrealistic performance goals are of concern. Most operators complained that they found it hard to balance between system red flags and the expectations of locals particularly during season shortages. In the regions of peri-urban Chennai, workers experience a plural burden; the pressure to follow the SCADA-instilled standards albeit dealing with emotionally electrified community work.

4 Results and Discussion

The comparative analysis of global case studies—from Dehradun, Barcelona and Singapore—provides a comprehensive view of the emotional demands imposed on frontline workers in smart water systems (SWS). While these systems enhance operational efficiency, they also produce new stressors and emotional burdens that are not evenly distributed across geographies, technologies, or worker demographics.

4.1 Global Smart Water Emotional Labour Situations

What all facilities faced is that the change towards SCADA-based and digitally monitored systems has increased the emotional regulation needs on the frontline staff. Dehradun, India, is a place where emotional exhaustion is enhanced with poor infrastructure, erratic weather, and network connection. When the failure of the system has external causes, employees bear the responsibility and end up being constantly frustrated and dissatisfied with the work.

On the contrary, the Aigues de Barcelona in Barcelona has a more mature technological environment. Nevertheless, emotional demands are there in the form of information overloading and active watchfulness. Frontline workers are supposed to overcome any emotional state in case of alarm fatigue, fear of false positive, and performance pressure.

Adding another spin of emotional labour, there exist AI-enabled systems to even more sophisticated levels regarding public expectations in the presence of diminished worker autonomy which occur in Singapore (PUB). The workers must align their professional identity to the systems that will supposedly make the decisions and take control away.

4.2 Emotions Invisibility and Blind Spots in Organization

The digital water management and labour practices in public utilities have transformed emotional, cognitive, and operational expectations of the frontline workforce. The results of the case studies show the subtle and much ignored emotional landscape in which these employees must manoeuvre. Although the benefits of digital infrastructure are considerable in terms of the derived operational enhancement, the human cost, as the aspect connected with emotions, is beyond substantial. The case studies show typical mental health issues due to algorithmic supervision, lack of training, and socio-cultural inconsistency between the digital systems and the standards of communities. Cognitive overload and particularly alert fatigue were especially high with real-time monitoring and in cases where accountability was linked to KPI measures.

Employees in the context of developing countries particularly India were subjected to other disadvantages such as low digital literacy, language barrier in interfaces and high communities' dependency. Emotional labour became more intense as technicians were supposed to play two roles of an engineer and a community mediator.

The JD-R model can assist in putting the resultant burn out caused by high demands (surveillance, being present in the moment) as combined with low resources (training, psychological support) into the right context. On the other hand, in Philadelphian where peer mentoring and union existed, job satisfaction was stationary.

Unlike such areas as healthcare or education, where emotional work is officially acknowledged and promoted, utility work is traditionally defined as physical and technical. Emotion demands are not taken into consideration in the HR policy or in performance review. This invisibility leads to under-identification of stress, burnout and job dissatisfaction. There is also the case of emotional effort, which might not be accounted, this is in the case of being composed under pressure, or being able to empathise with the angry customers, or being able to manage fear when intervening in an emergency.

4.3 Watching and Algorithmic Control

Smart systems not only enlarge the amount, but also the speed of work. The workforce will have to react instantly to automated notifications, whenever and wherever. The SCADA-based platforms and the IoT-based working environments are a successful way of turning the workplace to the personal space of employees giving them an expectation of extending their days in the workplaces. The performance dashboards, the logs of alerts, and the virtual KPIs of algorithmic management systems are the sources of daily reminders of the managerial control. This causes psychological pressure, especially in cases where the measurements do not reflect the complexity or the nature of the on-ground situations.

4.4 Role Conflict, Emotional Dissonance

The role conflict can also be seen in the context of workers in smart water systems: dealing with the need to be efficient with a system and with the emotional requirements of people. Institutional measures (time it takes to respond to leaks, reduction in downtimes) and emotional tasks (assuring nervous customers, appealing to the local populace) are dissonant because of the mismatch. In the long run, it is one of the reasons leading to emotional exhaustion and poor job satisfaction.

4.5 The Depersonalization and Cynical Risk

Depersonalization is a state of mind that workers adopt against their employers as an act of defending against emotional dissonance, unrecognized emotional work and lack of proper support mechanisms that may in the long run deteriorate into personal work breakdowns. This is especially so on the cases where operators report that they feel like they are part of a machine or data processors instead of being a part of the state. With the reduction in engagement, organization cynicism increases, which distracts morale and long-term retention.

Recommendations

To guarantee that the process of implementing smart water will be sustainable not only in the aspects of technology but also on an emotional level, the transitional efforts of the utilities should build comprehensive frameworks that can promote the emotional well-being of the workforce.

5 Training in Emotional Intelligence and Resilience

Such topics as emotional intelligence, stress management, and resilience should be added to professional development through regular workshops. Such sessions can assist employees to improve on how they perceive and manage their emotion to high-stakes situation.

5.1 *Technology Design with Human-Centred Approach*

The frontline workers should be involved in the planning and implementation process of smart system design. Dashboards and alert systems must be proved as usable, alert thresholds must be configurable, and operators must be consulted as to whether systems can aid but not slow their working practice.

5.2 *AI Assisted Alert Moderation and Prioritization*

To curb than alert fatigue, advanced analytics can be used to prioritize alerts according to their severity with non-critical alerts being filtered. With prioritized alerts, human attention will only be kept on those events that need it.

5.3 *Counselling and Peer Co-help Programmes*

Licensed counsellors and peer-based support groups are the ways which can make emotional discussions at the workplace normalized. Deliberately timed post-briefing of significant service adequacy lapses, or high-stress times enable staff members to decompress.

5.4 Reconfiguration of Shift Patterns and Recovery Intervals

Flexibility of shifts and the planned recovery time is very necessary in alleviating cognitive overload. Fatigue management policies should be observed by companies, treating mental rest as needed.

5.5 The Awareness of Emotional Work in HR Practices

Emotional labour measures must be part of the HR evaluation criteria. It is necessary to reward and recognize the workers who show a good level of emotional regulation, customer service, and teamwork in stressful situations. It values emotional labour since it is recognized during appraisals and promotions.

5.6 Policy Advocacy and Reform of the Institution

Policy makers and the public utility regulators should acknowledge emotional well-being as one of the pillars of sustainable urban infrastructure. Emotional health metrics and well-being audits ought to form the part of national utility standards.

6 Conclusion

With the rise of smart water solutions being implemented in the cities all over the world to support the changing urbanization conditions, the evolving needs of climate variation and the queries of resource optimization, the position of front-line utility workers also becomes more complicated. Such workers are not technicians anymore; they are managed systems, client service executives, data interpreters, and emotion buffers.

The future of efficient data-driven utility management is smart water systems, which also come at an emotional price. The frontline operators especially in the developing world such as India cannot be viewed no more than just passive routers in a digital network. When ignored, emotional labour can eventually eat through job satisfaction, leaving a systematically burnt-out condition and thus destroying the very fruits of the productive efficiency such systems were supposed to produce.

This chapter builds the argument in support of the sympathetic, allowance-based computer legislation where human welfare and technological inventiveness are given wide consideration. The possible future directions of research are to consider longitudinal impacts of algorithmic surveillance on emotional resilience and the evaluation of interventions that will be prepared collaboratively with frontline workers.

As shown in this chapter, with all the operational advantages that smart water technologies bring, there exist emotional pressures that can result in burnout, disengagement, and staff turnover unless the pressures are recognized. The concept of emotional labour in smart utilities should be perceived as a component of the new digital infrastructure.

Emotional well-being must be the focus of the holistic approach to the strategies of governance that are needed in digital water. This involves the redesigning of technology, institutional support mechanisms, evaluation practices in HR and frameworks of the policies. Future studies must look forward to probing the concept of emotional labour in other fields of infrastructure and see how the long-term effects of algorithmic surveillance are going to affect the identity and satisfaction rates of the worker.

Efficiency is not the only characteristic that smart water systems should have—they ought to be empathetic. Humanizing digital infrastructures starts with the acknowledgement and appreciation of the feelings of the people whose work makes them run.

References

Bakker, A. B., & Demerouti, E. (2007). The job demands-resources model: State of the art. *Journal of Managerial Psychology, 22*(3), 309–328.

Beker, B. A., & Kansal, M. L. (2022). Application of SCADA and IoT for efficient management of urban water supply system. In *International Conference Innovation in Smart and Sustainable Infrastructure* (pp. 657–674). Springer Nature Singapore.

Demerouti, E., & Bakker, A. B. (2011). The job demands-resources model: Challenges for future research. *SA Journal of Industrial Psychology, 37*(2), 01–09.

Demerouti, E., Bakker, A. B., Nachreiner, F., & Schaufeli, W. B. (2001). The job demands-resources model of burnout. *Journal of Applied Psychology, 86*(3), 499.

Frana, P. L. (2024). Assessing smart nation Singapore as an international model for AI responsibility. *International Journal on Responsibility, 7*(1), 2.

Grandey, A. A. (2000). Emotional regulation in the workplace: A new way to conceptualize emotional labor. *Journal of Occupational Health Psychology, 5*(1), 95.

Guy, C., Kaplan, F., Kopka, J., Selbig, J., & Hincha, D. K. (2008). Metabolomics of temperature stress. *Physiologia Plantarum, 132*(2), 220–235.

Hochschild, A. R. (2003). 1983: The managed heart. Commercialization of human feeling.

Johnston, H. (2021). *Success story: Embracing IoT & smart metering for a water resilient barcelona.* Utility Magazine.

Maslach, C., Schaufeli, W. B., & Leiter, M. P. (2001). Job burnout. *Annual Review of Psychology, 52*(2001), 397–422.

Morris, J. A., & Feldman, D. C. (1996). The dimensions, antecedents, and consequences of emotional labor. *Academy of Management Review, 21*(4), 986–1010.

Okoli, N. J., & Kabaso, B. (2024). Building a smart water city: IoT smart water technologies, applications, and future directions. *Water, 16*(4), 557. https://doi.org/10.3390/w16040557

Public Utilities Board Singapore Pan_Ju_Khuan@pub.gov.sg. (2016). Managing the water distribution network with a Smart Water Grid. *Smart Water*, *1*(1), 4.

Schaufeli, W. B., & Taris, T. W. (2013). A critical review of the job demands-resources model: Implications for improving work and health. In *Bridging occupational, organizational and public health: A transdisciplinary approach*, 43–68.

Steffelbauer, D. B., Blokker, E. J. M., Buchberger, S. G., Knobbe, A., & Abraham, E. (2021). Dynamic time warping clustering to discover socioeconomic characteristics in smart water meter data. *Journal of Water Resources Planning and Management, 147*(6), 04021026.

Tarafdar, M., Tu, Q., Ragu-Nathan, B. S., & Ragu-Nathan, T. S. (2007). The impact of technostress on role stress and productivity. *Journal of Management Information Systems, 24*(1), 301–328.

Young, M. M. (2020). Implementation of digital-era governance: The case of open data in US cities. *Public Administration Review, 80*(2), 305–315.

Google Earth Engine for Mapping Rooftop Rainwater Harvesting Potential Using Random Forest Classifier and Spectral Indices

Ramanarayan Sankriti, Saravanan Subbarayan, N. R. Dakshinamurthy, Sivaranjani Sivalingam, Niraimathi Janardhanam, and E. Arivoli

Abstract As per the 2030 Agenda related to sustainable development, countries worldwide are necessitated to put substantial efforts in improvising the quality of life of their citizens. One such step is to abate the water stress through rainwater harvesting. The main objective of the study is to estimate the Rooftop rainwater harvesting potential in the study area of Hyderabad city by using Gould and Nissen Formula (1999) in integration with Google Earth Engine (GEE) datasets. The Climate-Hazards-Group-InfraRed-Precipitation-with-Station-data (CHIRPS) dataset has been used for annual rainfall and Sentinel-2A image collection for settlement mapping. Built-up area indices that are utilized in the present study include band ratio for built-up areas (BRBA), new built-up index (NBI), normalized difference building index (NDBI) and the urban index (UI). These spectral indices are involved in the extraction of settlement areas. These spectral indices were thoroughly incorporated through a random forest supervised procedure. From the study, it can be concluded that Random Forest classifier is worthy enough to extract spatial distribution of settlements in Hyderabad region (India) with the accuracy metric kappa coefficient, exceeding 90%.

Keywords Hyderabad (Telangana) · Random Forest classifier · GEE · Rain water harvesting · CN grid · Spectral indices

R. Sankriti · N. R. Dakshinamurthy
Chaitanya Bharathi Institute of Technology (Autonomous), Hyderabad, India

R. Sankriti · S. Subbarayan (✉) · S. Sivalingam · N. Janardhanam · E. Arivoli
National Institute of Technology, Tiruchirappalli, Tiruchirappalli, India
e-mail: ssaravanan@nitt.edu

S. Pathak and M. Pandey (eds.), *Geospatial Technology and Advanced Instrumentation in Addressing Water Resources Planning and Management*, Water Informatics for Water Resource Management, https://doi.org/10.1007/978-981-92-1602-4_3

1 Introduction

Rooftop rainwater harvesting (RWH) has gained increasing importance as a sustainable water management strategy in urban environments. Identifying potential roof catchments for RWH requires precise spatial analysis and classification techniques. Google Earth Engine (GEE) provides a powerful platform for processing vast amounts of geospatial data, enabling efficient mapping of rooftop areas suitable for rainwater harvesting. The integration of machine learning techniques, such as Random Forest (RF) classification, enhances the accuracy of identifying potential RWH sites by using spectral information derived from satellite imagery. RF, a robust and flexible machine learning algorithm, uses spectral data from remote sensing sources (such as Landsat or Sentinel imagery) to classify rooftop surfaces based on their material properties, slope, and surrounding environment. Spectral information helps in distinguishing between different rooftop materials, vegetation, and other urban features, which directly affect the feasibility of rainwater collection. GEE's cloud-based platform offers extensive computational power to process large datasets, making it an ideal tool for urban planners and environmental engineers to map RWH potential efficiently. The application of Random Forest classification in GEE allows for accurate and scalable assessments of rooftop catchments, supporting decision-making for urban water conservation initiatives. By leveraging spectral data and machine learning, RWH mapping can provide detailed insights into the best locations for harvesting rainwater, contributing to sustainable urban development and resource management.

In the study by Lupia et al. (2017), which explored the potential of RWH for enhancing urban food production. The study focused on urban environments, highlighting the role of harvested rainwater in addressing water scarcity and promoting sustainable agriculture. The authors assessed RWH capacity using spatial analysis, considering factors like roof area and rainfall patterns. Their findings suggest that integrating RWH into urban planning could contribute significantly to food security and water conservation. This research underlines the importance of innovative water management solutions in urban agricultural practices.

In the study by Adugna et al. (2018) which assessed the potential for RWH in large public institutions, focusing on its contribution to sustainable water management. By analyzing rooftops of public buildings, the study evaluated the volume of rainwater that could be harvested and its potential to supplement water supply. The authors also examined the feasibility of RWH in alleviating urban water stress, particularly in regions facing water scarcity. Their findings suggest that public institutions can play a critical role in promoting RWH as an alternative water source, contributing to environmental sustainability and resource efficiency.

In the study by Villar-Navascués et al. (2020) which evaluated the potential of RWH using clustering analysis to categorize roof catchments based on their water-harvesting capabilities. The study incorporated variables such as roof size, material, and rainfall patterns to optimize the potential for collecting rainwater in urban areas. Their approach provided a more detailed understanding of how different roof types influence RWH efficiency. The findings emphasize the importance of tailored RWH strategies to maximize water collection, supporting urban sustainability efforts and contributing to water resource management.

In the study by Javed et al. (2021) which conducted a comprehensive review of spectral indices used in urban remote sensing, providing insights into their applications for monitoring and analyzing urban environments. The study explored various indices such as the Normalized Difference Built-up Index (NDBI) and Urban Index (UI), highlighting their effectiveness in detecting built-up areas, vegetation cover, and impervious surfaces. The authors emphasized the role of spectral indices in enhancing urban land cover classification and change detection, contributing to improved urban planning and management. This review serves as a valuable resource for researchers and practitioners in the field of urban remote sensing. The formulas of NDBI, UI, NBI and BRBA are taken from the above study.

In the study by Sun et al. (2019), which developed a machine learning-based classification system for urban built-up areas, utilizing multiple classifiers and diverse data sources. The study highlights the effectiveness of integrating various machine learning algorithms to enhance classification accuracy in urban environments. By leveraging a combination of spectral information and ancillary data, the authors demonstrated improved performance in identifying built-up areas, contributing valuable insights into the use of machine learning for urban remote sensing applications. This work underscores the potential of advanced classification techniques to support urban planning and management.

In the study by Ghosh et al. (2014), that explored the application of RF classification for urban landscape analysis by utilizing Landsat archives alongside ancillary data. The study emphasized the effectiveness of combining seasonal maps through decision-level fusion to enhance classification accuracy. By integrating multiple data sources, the authors successfully captured the dynamic nature of urban environments, demonstrating the Random Forest algorithm's robustness in distinguishing various land cover types. This research contributes to the growing body of literature on remote sensing techniques for urban studies, highlighting the potential of machine learning methods in facilitating informed urban planning and management.

In the study by Gupta and Panigrahy (2008) that investigated the spatio-temporal variation of runoff generation in India by employing remotely sensed inputs and the Soil Conservation Service Curve Number (SCS-CN) model. The study demonstrated the effectiveness of integrating remote sensing data with hydrological modeling to predict runoff patterns across different landscapes. By analyzing spatial and temporal variations, the authors highlighted the model's capability to inform water resource management and conservation strategies.

The objectives of the current study include: (1) To assess the potential of rooftop RWH using GEE and Sentinel-2A satellite imagery for large-scale urban analysis.(2)To implement RF classification for accurately identifying and mapping rooftop surfaces suitable for rainwater harvesting, using key spectral indices such as the BRBA, NBI, NDBI, and UI (3) To apply the Gould and Nissen formula to estimate the rainwater harvesting potential of identified rooftops, integrating hydrological modelling with spectral analysis. (4) To compute the Curve Number (CN) grid to assess runoff potential for different land cover types, providing a comprehensive understanding of how land use influences rainwater harvesting effectiveness.

2 Study Area

The current area under study, Hyderabad (India), as depicted in Fig. 1, experiences an average annual rainfall of approximately 800–1200 mm, predominantly during the time from June to September called the monsoon season, although rainfall varies by significant amounts from year to year, thus affecting water availability for RWH very much. The mean annual temperature varies between 23.5 and 31.5 °C, with summer highs exceeding 39.5 °C and milder winters averaging in between 14.5 and 24.5 °C. Vegetation in the region primarily consists of scrubland, mixed deciduous forests, and urban green spaces, although rapid urbanization has limited natural vegetation. The soil types are predominantly red and black soils, characterized by moderate to high water retention capacity, which can influence the effectiveness of rooftop rainwater harvesting systems. With a population of approximately 10 million, Hyderabad exhibits a high population density, particularly in urban areas, leading to increased demand for water resources. The demographic profile reflects a youthful population, with a significant portion under the age of 35, and a diverse economy that includes IT, manufacturing, and services, further influencing land use and water demand. Additionally, the city is situated at an average elevation of about 500 m above sea level, with a mix of residential, commercial, and industrial land uses resulting from rapid urban development, ultimately affecting natural water drainage and availability. The latitude and longitude limits of the Hyderabad region are approximately 17.2998° N to 17.4756° N and 78.3872° E to 78.6000° E. These coordinates encompass the urban and peri-urban areas of Hyderabad, serving as a spatial reference for various studies related to urban planning, water management, and environmental assessments.

3 Methodology

The methodology (Fig. 2) for mapping rooftop rainwater harvesting potential in the Hyderabad region begins with defining the study area and collecting relevant data, including Sentinel-2A satellite imagery, meteorological data (rainfall and temperature statistics), and demographic and land use information (vegetation types and soil types). Data preprocessing involves atmospheric correction and cloud masking of the imagery, ensuring that the cloud cover is less than 5% to maintain the quality of the data. Following this, key spectral indices are calculated, including the BRBA (which is calculated as a band ratio specifically designed to identify built-up areas), NBI, NDBI and UI. The RF classifier is then employed to produce rooftop terraces by using training samples prepared based on roof types and land cover. The model

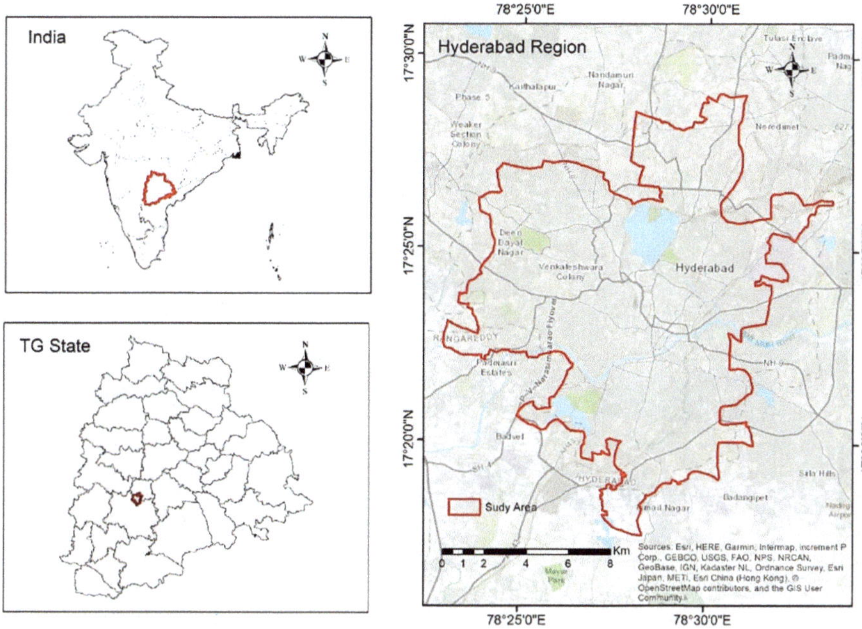

Fig. 1 Location map

is trained with the calculated spectral indices to classify rooftops suitable for rainwater harvesting, specifically identifying terraces designed for water collection. The accuracy of the Random Forest classifier is evaluated using the Kappa coefficient as a metric, which provides a measure of agreement between the classified results and information derived from the spectral indices while accounting for chance agreement. Subsequently, hydrological analysis is performed by applying the Gould and Nissen formula to estimate the rainwater harvesting potential for the classified rooftops, along with the computation of the CN grid to assess runoff potential for different land cover types. The methodology culminates in the creation of maps showing potential RWH sites and their associated metrics, followed by spatial analysis to evaluate the distribution and yield of suitable rooftops. Finally, the findings are interpreted in the context of urban water management and sustainability. The Gould and Nissen formula is given by:

$$\text{RWH} = P \times A \times \text{CC} \tag{1}$$

where the terms, RWH stand for RWH potential (measured in litres), 'P' stands for average annual rainfall (measured in mm), 'A' stands for the plan area of the rooftop (in square meters), 'CC' stands for runoff coefficient (a dimensionless factor accounting for losses during collection and storage, typically ranging from 0.5 to 0.9)

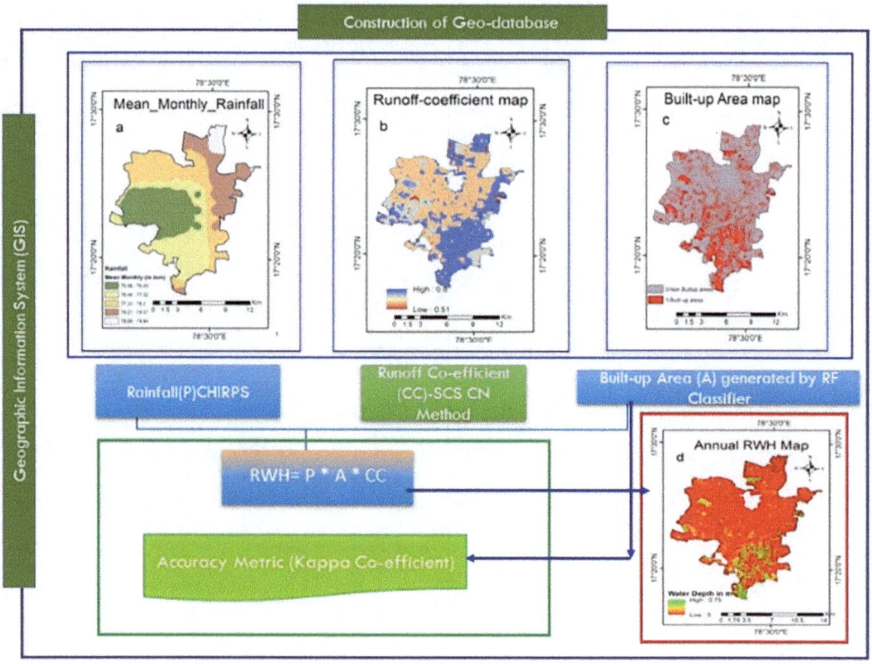

Fig. 2 Methodology flowchart

In the context of estimating RWH potential using the Gould and Nissen formula, the term '*P*' was calculated from the CHIRPS dataset in GEE. The process involves importing the CHIRPS dataset, defining the area of interest and filtering the data for the desired time frame (which is calendar year 2023–2024). By employing GEE functions, the mean precipitation for the specified area and period can be computed. This average rainfall value is then incorporated into the formula to effectively assess the rainwater harvesting potential of rooftops.

To estimate rooftop areas using a RF classifier, high-resolution satellite imagery, such as Sentinel-2, is utilized along with relevant spectral indices like the NDBI and UI. The process begins with data collection and feature extraction, followed by the preparation of a labelled training dataset that distinguishes rooftops from other land cover types. The RF model is then trained on this dataset to learn the spectral characteristics associated with rooftops. Once the model is utilized to classify the entire area of interest, the rooftop areas were calculated by summing the classified pixels and converting them into real-world measurements. Accuracy is assessed through metrics like the Kappa coefficient and overall accuracy, ensuring reliable results for urban planning and rainwater harvesting evaluations.

A runoff coefficient raster was generated using the HeC-GeoHMS tool in ArcGIS, by preparing the Sentinel-2 land use dataset (10 m resolution) and the Global Hydrologic Soil Groups (HYSOGs) dataset (250 m resolution). Resample the HYSOGs dataset to match the land use resolution. Load both datasets into ArcGIS, and use the HeC-GeoHMS extension to assign curve numbers based on land use classifications and hydrologic soil groups. Then, create a CN grid by combining these datasets. Convert the CN values into runoff coefficients in the resulting runoff coefficient raster.

4 Results and Discussion

The analysis utilized the NDBI with a range of 0.2–0.7 to effectively extract rooftop terraces from the urban landscape. The RF classifier, comprising 180 decision trees, demonstrated exceptional performance, achieving an overall accuracy and kappa co-efficient of 0.97 and 0.95, respectively, for mapping settlement areas. These results indicate a high level of agreement between the classified data and the actual land cover, confirming the classifier's robustness in distinguishing built-up regions. Table 1 shows the areal extent of built-up area in a map extracted from RF Classifier.

Fine-tuning of the RF model was performed by incorporating additional spectral indices, including the UI, BRBA, and the NBI. This inclusion enhanced the model's ability to capture the complexities of urban structures and improve classification precision. The diverse spectral information from these indices allowed the RF classifier to better differentiate between rooftops and other land cover types, leading to improved accuracy in built-up area mapping.

A confusion matrix was generated to assess the classifier's performance further, providing detailed insights into the classification outcomes. The matrix highlighted the model's strengths in correctly identifying built-up areas while also revealing any misclassifications. The high Kappa coefficient, in conjunction with the overall accuracy, reinforces the reliability of the RF classifier in urban mapping applications.

Overall, the integration of multiple spectral indices and the application of the RF classifier with 180 decision trees proved effective for accurately mapping rooftop terraces. This underscores its potential for urban planning and rainwater harvesting assessments. The findings suggest that using advanced remote sensing techniques can significantly enhance the understanding of urban environments, providing valuable data for sustainable water resource management.

Tables 2 and 3 refers to the components of a tertiary classification matrix in the general and specific cases in the current study respectively.

Table 1 Areal extent of LU classes resulted in RF classifier

LU/LC category	Area (in ha)	NDBI range	Suitability for roof top terraces
Water	1633	< 0	Non-built-up areas (water bodies)
Bareland/vegetation	12,225	0–0.2	Non-built-up areas (vegetation and Bare land)
Built-up	4387	0.2–0.7	Excellent (areas predominantly including roof top terraces)

Table 2 Elements of the tertiary confusion matrix in general

	Predicted P	Predicted Q	Predicted R
Actual P	TP (P)	FP ($P \rightarrow Q$)	FP ($P \rightarrow R$)
Actual Q	FN ($Q \rightarrow P$)	TP (Q)	FP ($Q \rightarrow R$)
Actual R	FN ($R \rightarrow P$)	FN ($R \rightarrow Q$)	TP (R)

Table 3 Confusion matrix pertaining to present study

	Predicted P	Predicted Q	Predicted R
Actual P (aquaculture pond)	32	0	0
Actual Q (non-aquaculture water)	0	22	0
Actual R (non-water area)	0	2	22

The runoff coefficient, calculated based on the derived CN values, ranged from a minimum of 0.51 to a maximum of 0.80. This range reflects the varying potential for runoff across different land uses and surface types in the study area, with higher coefficients indicating surfaces that are more likely to contribute to runoff. The results highlight the influence of urbanization on hydrological responses, emphasizing the importance of incorporating runoff coefficients in urban water management strategies. Figure 3 depicts spatial variation of parameters in computation of RWH potential.

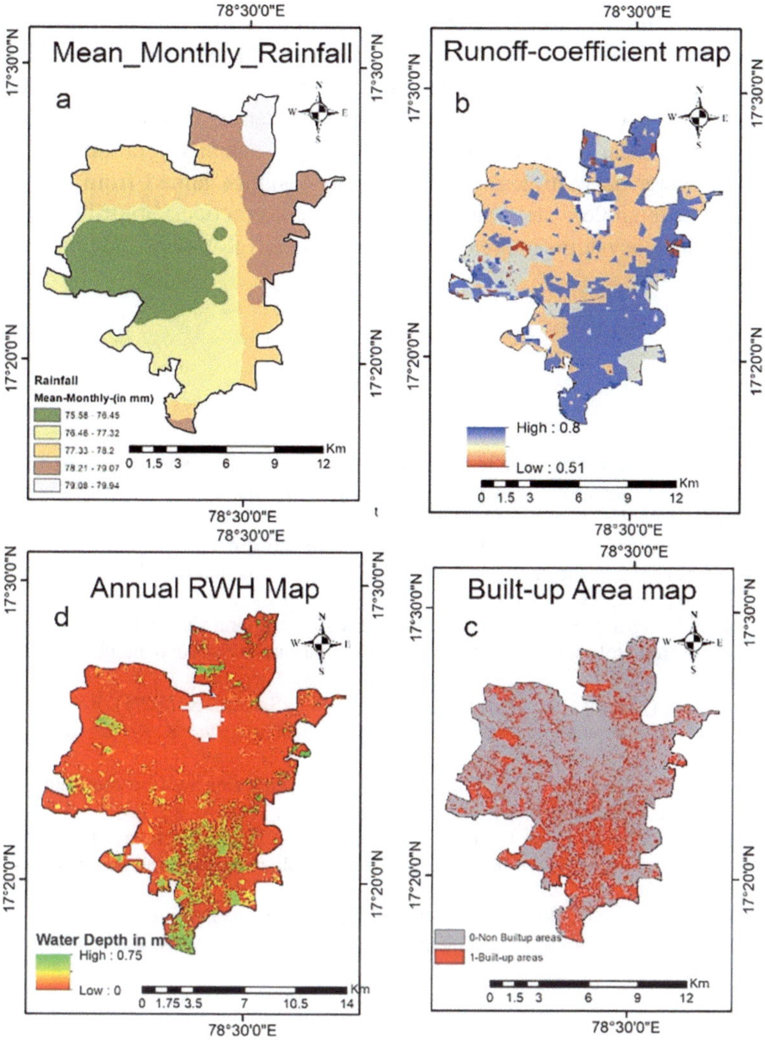

Fig. 3 Spatial variation of different parameters in Gould and Nissen formula and the annual RWH map

5 Conclusion

This study successfully demonstrated the effectiveness of using the RF classifier, consisting of 180 decision trees, in accurately mapping rooftop terraces within an urban environment. By leveraging spectral indices such as the NDBI, UI, BRBA, and NBI, the model achieved an impressive overall accuracy of 0.97 and a Kappa coefficient of 0.95 impressive overall accuracy of 0.97 and a Kappa coefficient of 0.95. Additionally, the analysis of runoff coefficients revealed a range from 0.51 to

0.80, highlighting the variability in runoff potential across different land uses and surfaces. These findings underscore the critical role of urbanization in influencing hydrological dynamics and emphasize the importance of incorporating runoff coefficients in urban water management strategies. Overall, this study illustrates the potential of advanced remote sensing techniques in enhancing urban planning and rooftop rainwater harvesting assessments. The insights gained from this research can inform sustainable water resource management practices, contributing to more resilient urban ecosystems in the face of increasing development and climate change challenges.

References

Adugna, D., Jensen, M. B., Lemma, B., & Gebrie, G. S. (2018). Assessing the potential for rooftop rainwater harvesting from large public institutions. *International Journal of Environmental Research and Public Health, 15*(2), 336.

Ghosh, A., Sharma, R., & Joshi, P. K. (2014). Random forest classification of urban landscape using Landsat archive and ancillary data: Combining seasonal maps with decision level fusion. *Applied Geography, 48*, 31–41.

Gupta, P. K., & Panigrahy, S. (2008). Predicting the spatio-temporal variation of run-off generation in India using remotely sensed input and Soil Conservation Service curve number model. *Current Science*, 1580–1587.

Javed, A., Cheng, Q., Peng, H., Altan, O., Li, Y., Ara, I., & Saleem, N. (2021). Review of spectral indices for urban remote sensing. *Photogrammetric Engineering & Remote Sensing, 87*(7), 513–524.

Lupia, F., Baiocchi, V., Lelo, K., & Pulighe, G. (2017). Exploring rooftop rainwater harvesting potential for food production in urban areas. *Agriculture, 7*(6), 46.

Sun, L., Tang, L., Shao, G., Qiu, Q., Lan, T., & Shao, J. (2019). A machine learning-based classification system for urban built-up areas using multiple classifiers and data sources. *Remote Sensing, 12*(1), 91.

Villar-Navascués, R., Pérez-Morales, A., & Gil-Guirado, S. (2020). Assessment of rainwater harvesting potential from roof catchments through clustering analysis. *Water, 12*(9), 2623.

Decadal Morphodynamic Evolution of the Sutlej River Basin: A Geospatial Perspective on Channel Shifts and Basin Response

Simranjeet Kaur and Shray Pathak

Abstract This study investigates the geomorphological dynamics and riverbank instability of the Satluj River along the Jalandhar boundary in Punjab using morphometric analysis and multitemporal satellite imagery from 2014 to 2024. Morphometric parameters derived from SRTM DEM reflect an elongated basin with moderate drainage texture and a low bifurcation ratio, indicating limited structural disturbance but high erosion potential. River channel delineation using Landsat 8 and Sentinel-2A imagery revealed substantial lateral migration, with the left bank undergoing maximum erosion of − 366.47 m (Reach 11) and the right bank showing peak deposition of + 238.77 m (Reach 13). The central reaches (7–13) emerged as critically vulnerable zones, with simultaneous erosion and deposition reflecting unstable channel morphology. These changes are exacerbated by post-2014 flood events and anthropogenic activities such as sand mining and unregulated land use. The findings highlight that the Satluj River is facing accelerated bank erosion, loss of riparian stability, and degradation of adjacent land—necessitating urgent developmental interventions, including sustainable riverbank protection, habitat restoration, and long-term conservation planning.

Keywords Satluj River · Bank erosion · Morphometric analysis · Lateral shifts · Remote sensing

S. Kaur (✉)
Department of Geography, Panjab University, Chandigarh, India
e-mail: simranjeetpunia092@gmail.com

S. Pathak
Department of Civil Engineering, Indian Institute of Technology Ropar, Rupnagar, India

S. Pathak and M. Pandey (eds.), *Geospatial Technology and Advanced Instrumentation in Addressing Water Resources Planning and Management*, Water Informatics for Water Resource Management, https://doi.org/10.1007/978-981-92-1602-4_4

1 Introduction

River morphology is a branch of geomorphology that focuses on the shape and structure of streams and their surrounding landscapes. It studies the spatial and temporal changes brought over them by processes of erosion, sediment transport, and deposition caused by flowing water (Garde, 2006). Understanding morphological changes is crucial for water resource management, flood control, agricultural planning, and the environmental conservation of any region. One of the most important tools in understanding River morphology is the morphometric survey of a basin, which allows one to establish evaluation parameters based on the behavior of the hydrological system of the basin area (Philip et al., 2014). Morphometric descriptors help researchers understand and compare the way river basins behave by looking at their shape and structure. Elements like geology, terrain, and climate play a key role in shaping water flow. Studies carried out by Horton (1945), Strahler (1957) and others show that over time, rivers naturally adjust their form to balance water and sediment movement (Ali et al., 2012). Morphometric analysis involves the quantitative measurement and mathematical analysis of landform configuration, including the geometry of drainage basins and channel networks (Horton, 1945). This approach provides valuable insights into the hydrological behavior, erosional processes, and geological controls of river basins through the systematic measurement of various morphometric parameters (Strahler, 1957).

River morphology is controlled by a complex interplay of multiple factors. Structural and land use changes such as dam construction, channel improvement, detention storage, forestry, agriculture, mining, and urbanization affect rivers to a great extent; this results in bank erosion and poor water quality (Stephens et al., 2021). Bank erosion is a complex geomorphic process that represents one of the most significant challenges toward river management in India, and the world at large (Goswami et al., 2024). Riverbank erosion is a natural geomorphic process that occurs across all streams, involving adjustments in channel size and shape to accommodate the discharge and sediment load delivered from the drainage basin. The deposition and erosion of sediments within the river significantly influence the cross-sectional area, slope, flow, velocity, and discharge capacity of the channel. Consequently, these geomorphic alterations can lead to river overflows, often resulting in flooding in adjacent areas (Avtar et al., 2011). Bank erosion promotes riparian vegetation succession and creates dynamic habitats crucial for aquatic and riparian plants and animals, but river managers and policymakers generally regard bank erosion as a threat (Florsheim et al., 2008). It is considered a threat because it leads to widespread damage and substantial economic losses. In India, this risk is amplified by a combination of dense populations, socio-economic vulnerabilities, and diverse physiographic conditions, making several regions particularly susceptible to erosion-related hazards (Bhuyan et al., 2025). This emphasizes the need for comprehensive studies to address emerging erosion factors in the context of increasing human activities and climate change. Recent advances in Geospatial technologies such as remote sensing and GIS offer efficient, cost-effective tools for monitoring river geomorphodynamics. They enable spatiotemporal analysis of river and floodplain changes using synoptic data. Unlike traditional field methods, these tools provide rapid, large-scale insights into historical and ongoing morphological shifts (Langat et al., 2019). Thematic maps created using Landsat TM and SRTM DEM help categorize geomorphological and hydrological features to assess inundation and erosion intensity. This analysis identified zones prone to erosion and flooding at the Ken-Betwa River linking site in India (Rozo et al., 2014).

The Satluj River basin, one of the largest river systems in the Indian subcontinent, presents a complex morphometric framework due to its vast extent and diverse physiographic characteristics. Originating from the Tibetan Plateau and flowing through the Himalayas into the Punjab plains, the basin comprises of varied topographic conditions ranging from high-altitude glaciated terrain to low-lying alluvial plains (Bhutiyani et al., 2008). This diversity makes the Sutlej basin an ideal case study for comprehensive morphometric analysis. The lower Satluj River basin—encompassing ten districts of Punjab—has long struggled with soil erosion, a persistent and silent threat. Once a thriving mosaic of fish, amphibians, turtles, and otters, the region's biodiversity is now gradually fading (Deshmukh, 2025). Thus, it is crucial for engineers working in river environments to identify and anticipate potential channel instabilities. An overview of geomorphic erosion processes is essential in interpreting the physical mechanisms behind channel changes, as it supports better decision-making during design and planning (Trent & Brown, 1984).

This study aims to delineate the geomorphological changes in River Satluj, prepare a drainage map of its basin, and compute the morphometric parameters. Further, it aims to calculate the lateral bank shifts in the banks of the Satluj River from 2014 to 2024, while analyzing its natural and anthropogenic driving factors. On this basis, erosional and depositional zones are to be estimated and vulnerable zones are to be identified.

2 Study Area

The study area focuses on a section of the Satluj River in Punjab, along the boundary of the Jalandhar district. The total length of the Sutlej River along the boundary of Jalandhar is approximately 108 km. The Satluj River originates from the western part of Lake Mansarovar in Tibet and is a major tributary of the Indus River. It enters India near the Shipki La Pass, flowing through several Himalayan valleys before merging with the Beas River in Punjab. The Beas then converges with the Chenab in Pakistan, eventually joining the Indus. In Indian Punjab, the Satluj enters near Nangal, flows through Ropar and Ludhiana, and meets the Beas at Harike before crossing into Pakistan (Setia et al., 2020). Jalandhar, the north-eastern district of Punjab State located between 30°59′ and 31°37′N latitudes and 75°04′ and 75°57′E longitudes, is considered the study area (Fig. 1) having a total geographical area of around 2662 m^2. The mean elevation of the study area varies between 184 and

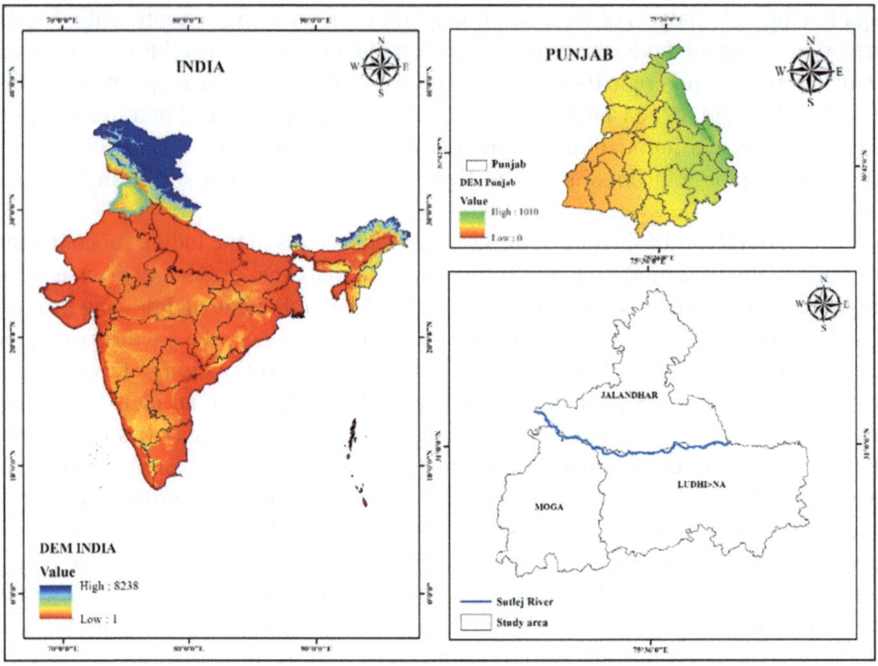

Fig. 1 Study area

294 m [extracted from SRTM (Shuttle Radar Topography Mission) DEM] (Singh et al., 2015).

The lower Satluj River basin faces grave challenges from soil erosion and sedimentation, which compromise its water retention capacity and degrade its downstream water quality. The continuous loss of topsoil diminishes agricultural productivity and heightens the risk of landslides. To restore and protect the region, it is crucial to implement erosion control measures such as terracing, vegetative barriers, and sustainable land management practices; thus, understanding the river's channel patterns is imperative in safeguarding the region (Gupta, 2025). To achieve the objectives of the study, multi-temporal satellite images from different sensors were utilized. Landsat 8 imagery from the year 2014, with a spatial resolution of 30 m, was acquired from the USGS Earth Explorer. For more recent and higher-resolution observations, Sentinel-2A imagery from year 2024 was sourced from the Copernicus Data Space Ecosystem, offering a finer spatial resolution of 10 m.

3 Methodology

To accomplish the objectives of the present study, the methodology is broadly divided into Stream profile analysis (Fig. 2) and image processing of the satellite data (Fig. 4).

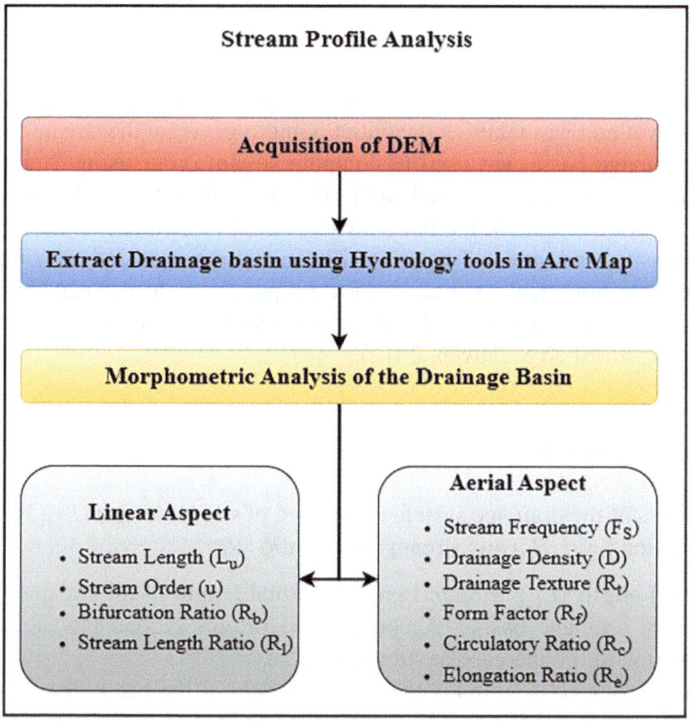

Fig. 2 Methodological framework of stream profile analysis

3.1 Stream Profile Analysis

The Digital Elevation Model (DEM) from the Shuttle Radar Topography Mission (SRTM) was obtained from the USGS Earth Explorer for the study area covering the Sutlej River stretch along the boundary of Jalandhar, Punjab. Initially projected in WGS 1984, the DEM was reprojected to the Universal Transverse Mercator (UTM) coordinate system, Zone 43, to align with the spatial reference requirements of the study. The study area was then delineated using the "Extract by Mask" tool from the Spatial Analyst toolbox, which isolated the raster cells that fall within the defined mask area, ensuring focused analysis on the specified region.

In ArcMap, the Spatial Analyst Hydrology tools were then used to delineate the drainage basin. This process involved several key steps:

1. **Fill Sinks**: This step modified the DEM to remove any depressions or pits that could have disrupted the flow direction calculations.
2. **Flow Direction**: Determined the direction of water flow from each cell to its steepest downslope neighbor.
3. **Flow Accumulation**: Calculated the accumulated flow as it traversed the landscape, identifying the drainage pattern.

4. **Watershed**: Defined the boundaries of the drainage basin based on the flow accumulation and direction data.

Once the drainage basin was delineated, morphometric analysis was conducted. The Harike wetland was taken as the outlet point to generate the stream profile. The resulting drainage basin was used to compute stream order using Strahler (1964) nomenclature system and stream length. This allowed the calculation of various Linear and Aerial aspects of the basin's characteristics, such as its area, perimeter, drainage density, and stream frequency. These metrics provided valuable insights into the geomorphological and hydrological properties of the drainage basin, aiding in understanding its behavior and supporting effective resource management and planning. (Saeedrashed & Guven, 2012).

3.1.1 Linear Aspects

Linear aspects of the drainage system comprised of stream length (L_u), stream order (U), bifurcation ratio (R_b) and stream length ratio (R_L).

a. **Stream Length (L_U)**: Stream Length is "total length of all streams of order u in a given drainage basin". According to Horton's second law, stream length decreases with the increase in stream order.
b. **Stream Order (U)**: To index the size and scale of the basin, the classification of stream order needed to be completed. Head water streams that receive no tributary are first order stream i.e. originates at a source. When two streams of order u join, a stream of order $u + 1$ is formed. Figure 3 shows the Strahler stream order.
c. **Bifurcation Ratio (R_b)**: Bifurcation Ratio is estimated as "the ratio of number of streams of lower order u to number of streams of next higher order $u + 1$".

$$R_b = \frac{N_u}{N_u + 1} \tag{1}$$

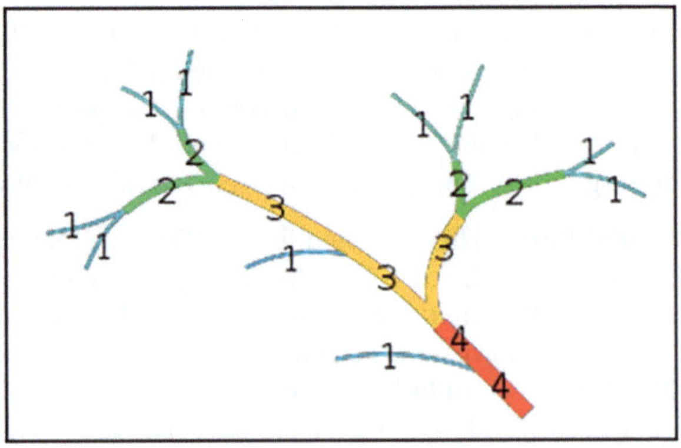

Fig. 3 Strahler stream order

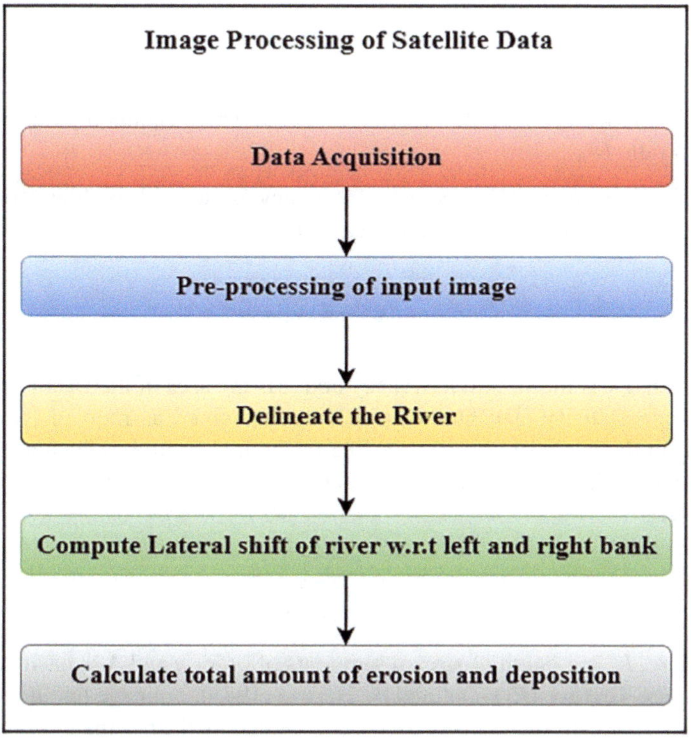

Fig. 4 Methodological framework of the image processing

where, R_b is the bifurcation ratio; N_u is the total no. of stream segments of order 'u'; $N_u + 1$ is the number of segments of the next higher order and R_{bm} is the average of bifurcation ratios of all orders.

Bifurcation ratios are related to the structural control on the drainage. A higher value of bifurcation ratio indicates strong strong structural control in drainage development, while a lower value shows less impact of structures on the drainage basin.

d. **Stream Length Ratio (R_L)**: Stream Length Ratio is "the ratio of mean stream segment of order u to that of order $u - 1$". Stream length ratio increases from lower to higher stream order.

$$R_l = \frac{L_u}{L_{u-1}} \tag{2}$$

where, R_l is the stream length ratio; L_u and L_{u-1} is the total stream length of order u and $u - 1$.

3.1.2 Aerial Aspects

Aerial aspects of the drainage basin comprise of stream frequency (F_S), drainage density (D), Drainage Texture (R_t), form factor (R_f), circulatory ratio (R_c) and elongation ratio (R_e).

a. **Stream Frequency (F_S)**: Stream frequency is defined as "the ratio of total number of streams per unit basin area" (Horton, 1933).

$$F_S = \frac{\sum_{u=1}^{n} N_u}{A} \tag{3}$$

where, F_S is the stream frequency and A is the area of the Basin in km^2.

b. **Drainage Density (D)**: Drainage density is defined as ratio of total channel length in the basin of all orders to the basin area Factors influencing drainage density include rock resistance to weathering, permeability, vegetation cover, and climate.

$$D = \frac{\sum_{u=1}^{n} L_u}{A} \tag{4}$$

where, L_u = Length of stream segments of order u and A = basin area (km^2)

c. **Drainage Texture (R_t)**: Defined by Horton (1945), drainage texture is the total number of stream segments of all orders divided by the perimeter of the basin. It is primarily governed by the infiltration capacity of the surface and also integrates drainage density and stream frequency.

$$R_t = \frac{\sum_{u=1}^{n} N_u}{P} \tag{5}$$

where, R_t is the drainage texture; P is the basin perimeter (km)

d. **Form Factor (R_f)**: Form factor is the ratio of basin area to the square of its length (Horton, 1933). A lower form factor indicates a more elongated basin, which typically has a longer runoff duration.

$$R_f = \frac{A}{{L_b}^2} \tag{6}$$

where, R_f is the form factor; and L_b is the basin length.

e. **Elongation Ratio (R_e)**: The elongation ratio measures the shape of the basin, calculated as the ratio of the diameter of a circle with the same area as the basin to the basin's maximum length. Values range from 0.6 to 1.0, where values close to 1.0 indicate a more circular basin, which is generally more efficient in runoff response.

$$R_e = \frac{D_c}{L_b} \tag{7}$$

where, R_e is the elongation ratio and D_c is the diameter of the circle.

f. **Circularity Ratio (R_C)**: Circularity ratio is the ratio of the basin area to the area of a circle having the same perimeter. It is influenced by the basin's slope, relief, geological structure, stream frequency, and land use/cover.

$$R_c = \frac{4\pi A}{P^2} \tag{8}$$

where, R_C is the Circularity Ratio.

3.2 Image Processing of Satellite Data

To prepare the satellite imagery for analysis, all datasets from 2014, and 2024 were reprojected to the Universal Transverse Mercator (UTM) coordinate system, WGS 84 Zone 43, to ensure spatial alignment with other geospatial layers. A False Color Composite (FCC) was then created by assigning suitable spectral bands to the red, green, and blue channels. This enhanced the visual interpretation of various land features, particularly highlighting water bodies, which exhibit strong reflectance differences in the Near-Infrared (NIR) region. Then, the river banks were delineated (Fig. 4).

Delineating a river involves precisely mapping its course, boundaries, and associated features such as tributaries, floodplains, and adjacent landscapes (Sarkar et al., 2012). This process typically employs a combination of field surveys, remote sensing, and Geographic Information Systems (GIS) to capture and analyze spatial data. The primary objective is to develop an accurate and detailed representation of the river system, which is vital for informed water resource management, environmental conservation, and urban development planning. River delineation also aids in understanding hydrological patterns, identifying flood-prone areas, and assessing the ecological importance of riparian zones (Sarma & Phukan, 2004). To delineate the river channels, individual shapefiles were created for the left and right banks for each of the study years 2014 and 2024—using the ArcCatalog module in ArcMap 10.4. Manual digitization of the riverbanks was carried out for all three years using the Editor tool. The 2001 bank lines were designated as the baseline for assessing river bank displacement over time. Equidistant points were generated at 5 km intervals along both the left and right banks of the 2014 shapefile. From these points, perpendicular transects were drawn to intersect the corresponding bank lines of 2024, ensuring consistent positional accuracy across datasets. Points of intersection between the transects and the bank lines were then marked, and the displacement distance for each bank was calculated using the *Near* tool, with the 2014 points serving as input. A positive (+) sign was used to indicate inward movement (erosion), while a negative (–) sign denoted outward movement (deposition).

4 Results and Discussion

4.1 Stream Profile Analysis

Flow direction, as shown in Fig. 5, depends on fill sinks by which the neighborhood terrain is levelled. Values from 1 to 128 show the direction of flow. Stream begins to form, as water flows downstream. Flow accumulation addresses the cell by counting the number of upstream cells which flow to a given cell. Figure 6 shows flow accumulation. The extracted stream network of the Satluj basin is shown in Fig. 7.

From Strahler stream ordering scheme, the total number of segments of each order in the Satluj drainage basin have been computed, and are given in Table 1.

The stream order data revealed a typical dendritic drainage pattern dominated by lower-order streams. The 1st order streams were the most abundant, with 1778 segments and a combined length of approximately 2,443,059 m. This number steadily decreased with increasing stream order, reflecting Horton's law of stream numbers and lengths. The 2nd and 3rd order streams also maintained substantial presence, indicating a dense and well-connected drainage network. As the stream order progressed from the 4th to the 6th, both the number of segments and their cumulative length dropped significantly, suggesting channel merging and increased

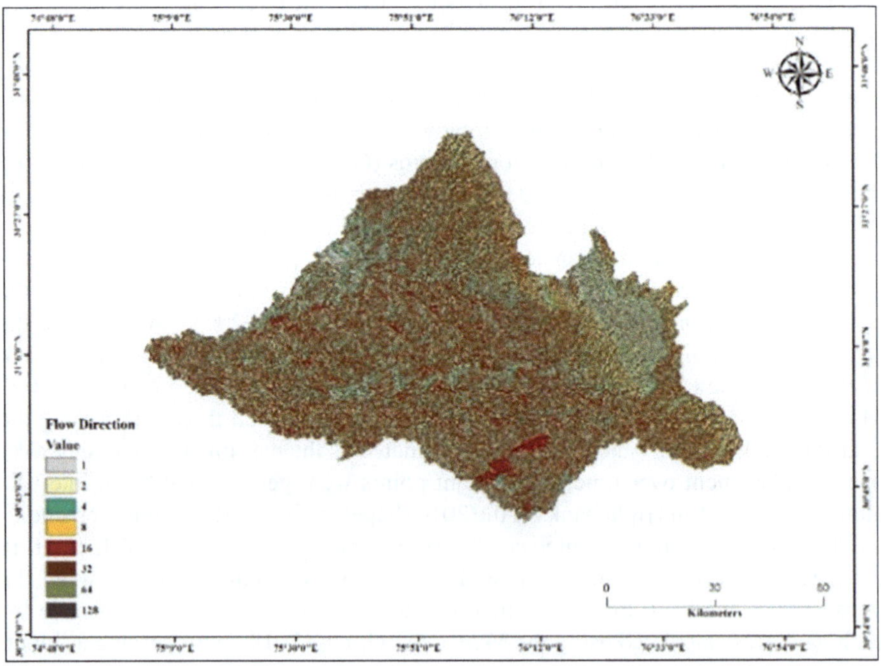

Fig. 5 Flow direction

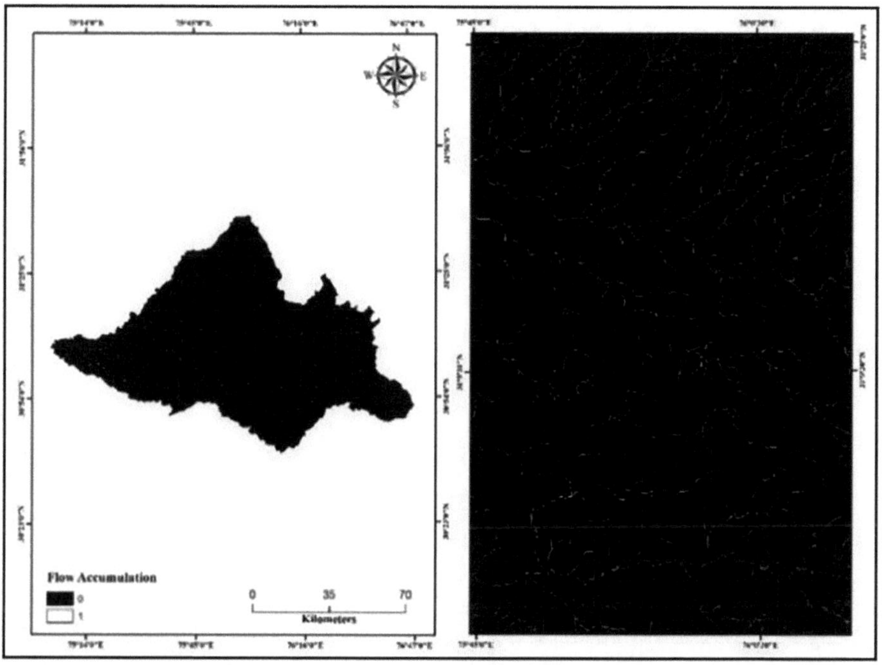

Fig. 6 Flow accumulation

discharge capacity downstream. The presence of 6th order streams signified a moderately mature drainage basin capable of handling substantial runoff. This distribution indicated a landscape with active surface flow, likely influenced by varied topography and possibly erodible soils, especially in upstream areas. The stream hierarchy reflected a hydrologically active basin with implications for sediment transport, erosion potential, and watershed management.

From the drainage network obtained through the processing of DEM, the basin characteristics were calculated (Table 2) and with the help of the morphometric formulas, the geomorphological parameters (Aerial and Linear Aspects) were computed, as given in Table 3. The morphometric parameters of the basin highlight its scale, complexity, and hydrological behavior. Covering a total area of 8252 km^2 with a perimeter of 734 km, the basin exhibited a substantial drainage region. The basin length of 165.22 km suggested an elongated form, which typically indicates a longer concentration time for runoff. With a dense stream network comprising 3633 streams and a cumulative stream length of 4945.86 km, the drainage system was highly dissected. This density reflected effective drainage efficiency and potential for quick hydrological response to rainfall events. Such characteristics are vital for understanding flood potential, sediment transport, and water resource planning within the basin. The lower Mean Bifurcation Ratio of 1.76 exhibited the characteristics of structurally less disturbed watersheds without any distortion in drainage pattern. Further, the elongation ratio value in the study area was 0.62. This indicated

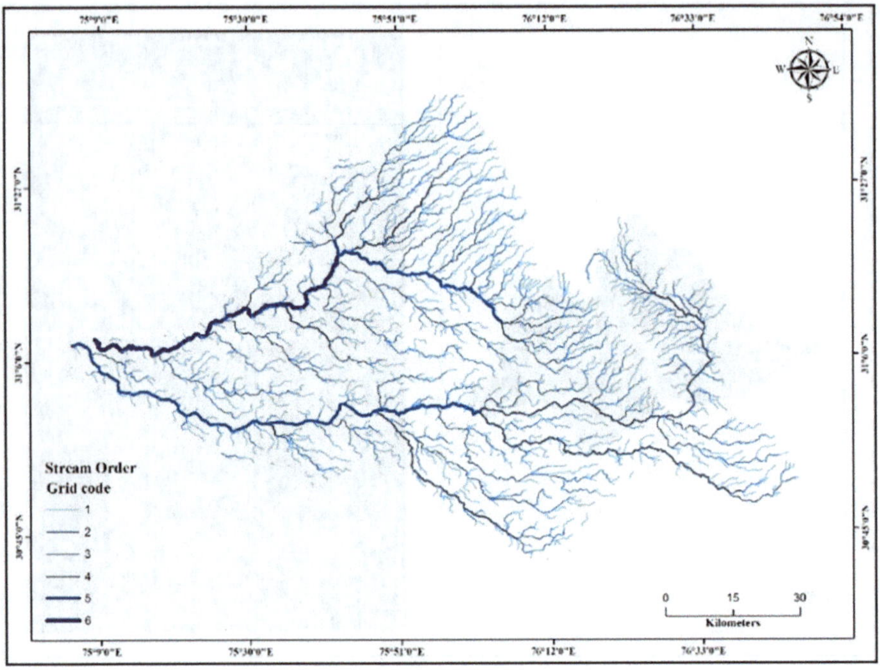

Fig. 7 Stream order of Satluj River Basin

Table 1 Classification of various stream orders

S. No	Stream order	Total No. of segments	Total length (m)
1	1st	1778	2,443,059
2	2nd	837	1,256,052
3	3rd	494	684,223
4	4th	224	313,577
5	5th	183	168,744
6	6th	117	80,205

that basin can be generally associated with strong relief and steep ground slope. The Drainage Density value of 0.599 indicated that the region is humid. Both drainage density and stream frequency measure the texture of drainage set, which is 4.95 i.e., moderate texture. The smaller value of form factor 0.30 and Circulatory ratio value 0.19 indicated that the basin is elongated in shape.

Table 2 Satluj River Basin characteristics

S. No	Parameters	Value
1	Basin area	8252 km^2
2	Basin perimeter	734 km
3	Basin length	165.22 km
4	Total stream length	4945.86 km
5	Total no. of streams	3633

Table 3 Linear and aerial aspects of Satluj River Basin (Kaur et al., 2025)

S. No	Linear aspects		Aerial aspects	
	Parameter	Value	Parameter	Value
1	Stream length	4945.86 km	Drainage density	0.599 km/km^2
2	Stream order	6	Stream frequency	0.44 streams/km^2
3	Mean bifurcation ratio	1.76	Circularity ratio	0.19
4	Mean stream length ratio	0.89	Form factor	0.30
5	–	–	Elongation ratio	0.62
6	–	–	Drainage texture	4.95

4.2 Analysis of Delineated River Banks

The delineation of the river channel was performed using high-resolution multispectral satellite imagery, specifically from Landsat 8, and Sentinel-2A. Pre-processing steps included radiometric correction, geometric correction, and cloud masking to ensure data quality and temporal consistency. Figure 8 illustrates the delineated Satluj River channel for the years 2014 and 2024. Figure 9 shows the overlap of the delineated river in the years 2014 and 2024 which shows how the Satluj River's course has shifted between 2014 and 2024. The channel has migrated laterally at several locations, with noticeable changes in its meandering pattern. These shifts suggest active fluvial processes such as erosion, deposition, and channel realignment over the years.

The Sutlej River demonstrated marked lateral instability between 2014 and 2024, with significant bank shifting driven by both natural and anthropogenic factors, as shown in Table 4. The left bank experienced maximum erosion of – 366.47 m in Reach 11, while the highest outward shift (deposition) was + 108.67 m in Reach 4. The right bank showed extreme variability, with the greatest erosion in Reach 7 (– 232.48 m) and maximum deposition in Reach 13 (+ 238.77 m). The central reaches (7–13) were identified as highly vulnerable zones, facing simultaneous left bank retreat and right bank expansion—suggesting active channel migration, meander shifting, and a risk of Bankline failure. These patterns were likely amplified by the post-2014 flood events, which may have increased discharge, sediment load, and erosive force, accelerating morphological changes. Combined with intensive sand mining

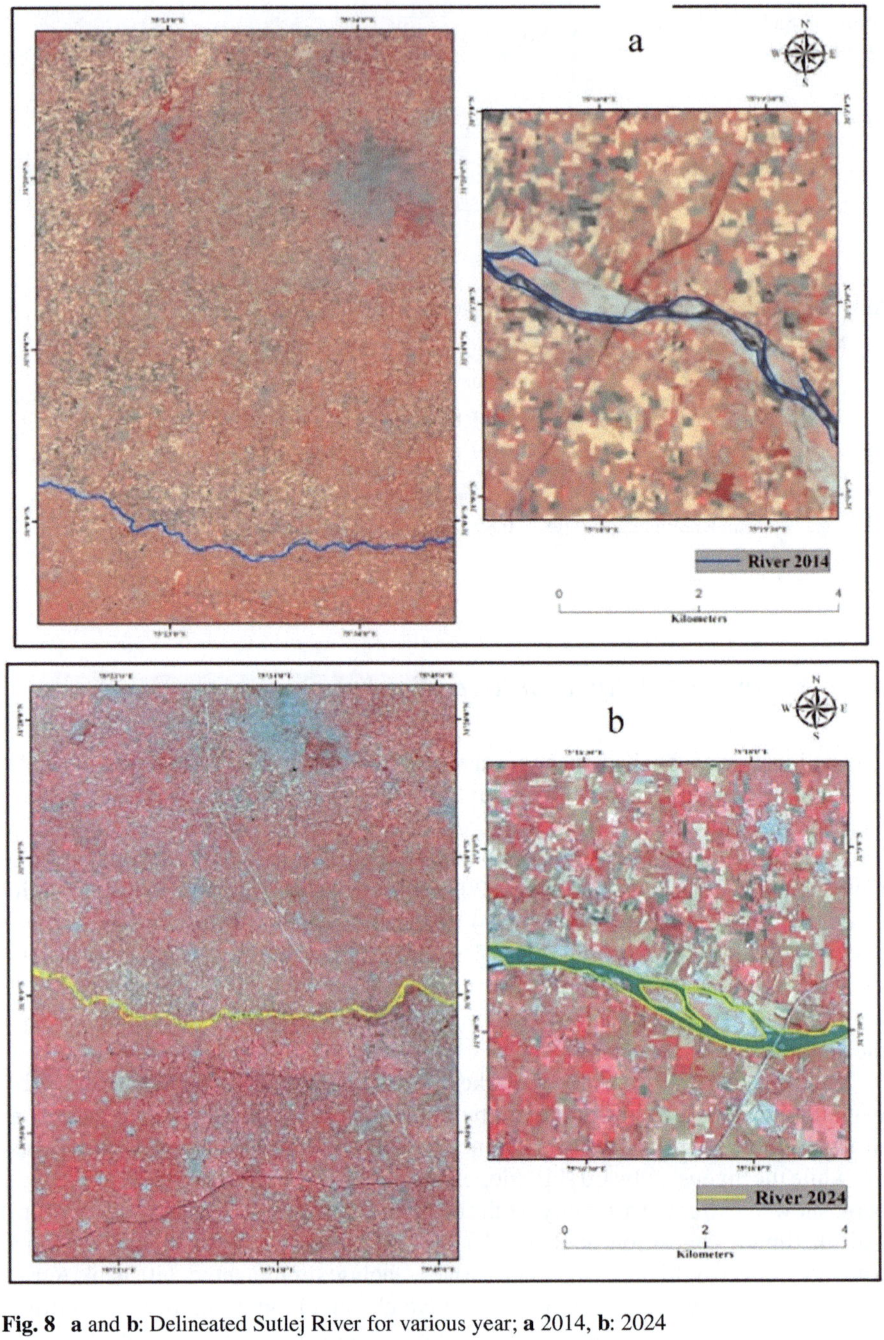

Fig. 8 **a** and **b**: Delineated Sutlej River for various year; **a** 2014, **b**: 2024

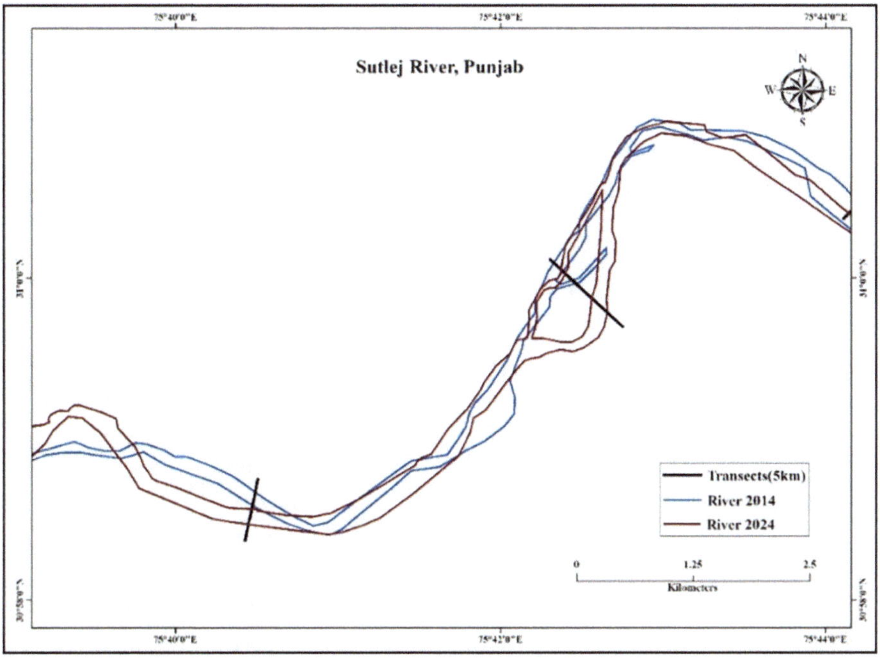

Fig. 9 Overlap of the delineated river, 2014 and 2024

and reduced vegetative cover, these floods have likely worsened channel instability, demanding urgent attention for riverbank protection and sustainable management (Fig. 10).

Table 4 Lateral shifting of left and right bank with respect to the year 2014

Reach	Distance (km)	Left bank (m)	Right bank (m)
1	5	– 72.37	50.45
2	5	24.11	83.81
3	5	28.05	– 20.57
4	5	108.67	– 48.59
5	5	94.69	– 216.14
6	5	– 40.41	– 49.27
7	5	– 146.27	– 232.48
8	5	– 227.63	124.85
9	5	– 205.9	151.75
10	5	– 219.68	186.85
11	5	– 366.47	231.9
12	5	– 265.12	87.7
13	5	– 193.24	238.77
14	5	– 116.05	34.42
15	5	– 153.73	75.84
16	5	– 118.56	– 102.66
17	5	– 33.44	– 28.13
18	5	40.47	– 199.7
19	5	13.17	– 4.49
20	5	48.02	61.92
21	5	– 10.45	– 103.31
22	5	– 46.01	33.02
23	5	– 16.27	30.99

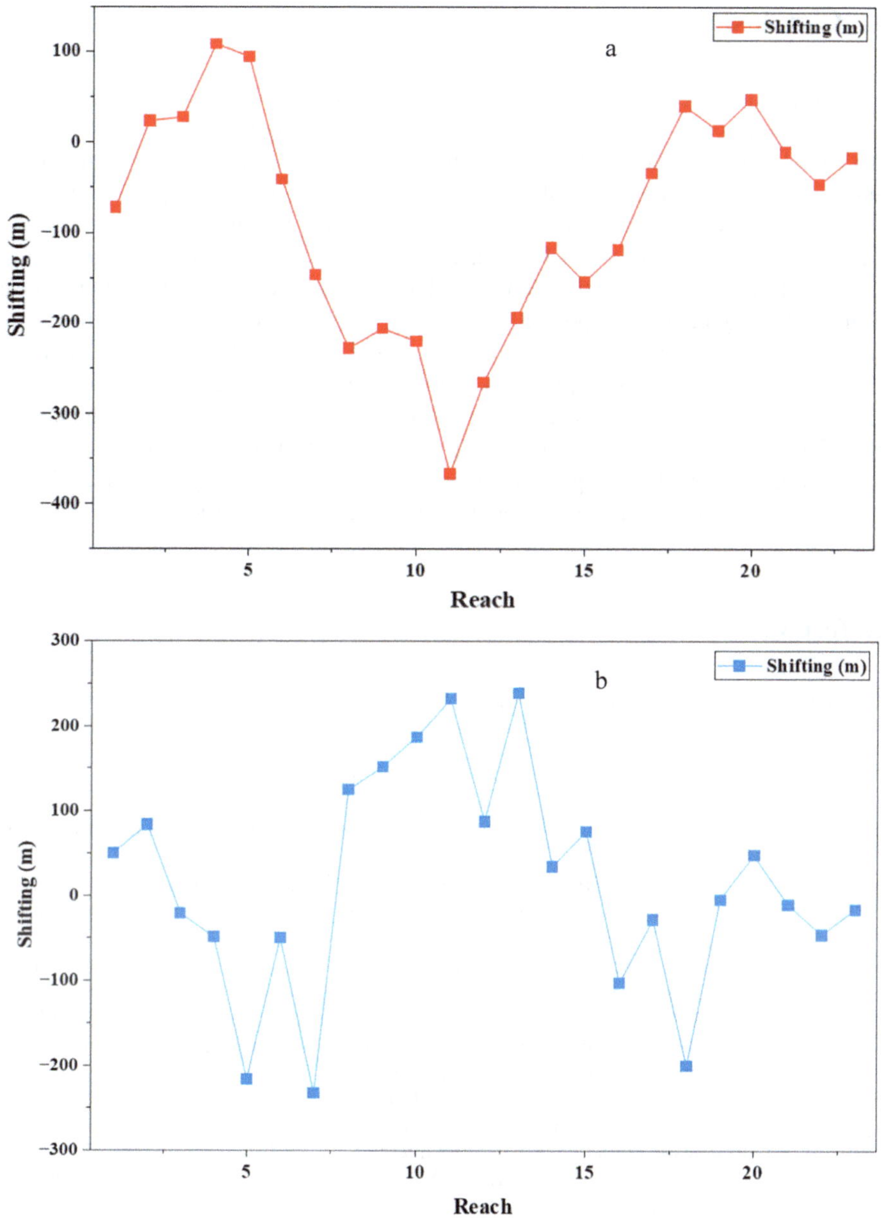

Fig. 10 **a** and **b**: Variation of shift at each chainage; **a** left bank, **b** right bank

5 Conclusion

This study underscores the dynamic and fragile state of the Satluj River, which has been undergoing intensifying lateral shifts and morphological changes over the last decade. Morphometric analysis confirms a moderately dissected yet structurally undisturbed basin, with potential for rapid runoff and sediment transport. The river's channel migration, particularly in Reaches 7–13, reveals a disturbing pattern of severe left bank erosion (up to – 366.47 m) and significant right bank accretion (up to + 238.77 m), signalling channel widening and instability. These erosional patterns have been further aggravated by post-2014 flooding, which introduced excess discharge and sediment, and by unchecked human activities like sand mining and land-use change. As a result, the Sutlej River is increasingly vulnerable, threatening agricultural land, local settlements, and ecosystem health. This study reinforces the urgent need for developmental action, including erosion control structures, riparian buffer restoration, regulated mining, and integrated river basin management policies to safeguard the long-term ecological and hydrological health of the region.

References

Ali, S. N., Bali, R., Agarwal, K. K., Rastogi, S. K., & Krishna, K. (2012). Drainage morphometry of Himalayan glacio-fluvial basin: Hydrologic and neotectonics implications. *Environmental Earth Sciences, 66*(4), 1163–1174.

Avtar, R., Singh, C. K., Shashtri, S., & Gupta, A. (2011). Identification of erosional and inundation hazard zones in Ken-Betwa river linking area, India, using remote sensing and GIS. *Environmental Monitoring and Assessment, 182*(1–4), 341–360. https://doi.org/10.1007/s10661-011-1880-6

Bhutiyani, M. R., Kale, V. S., & Pawar, N. J. (2008). Changing streamflow patterns in the rivers of northwestern Himalaya: Implications of global warming in the 20th century. *Current Science, 95*(5), 618–626.

Bhuyan, N., Sajjad, H., Rahaman, M. H., & Ahmad, F. (2025). Riverbank erosion induced vulnerability in India: A review for future research framework. *Natural Hazards, 121*, 1–30.

Deshmukh, A. (2025). Spotlight on erosion: Prioritizing the watersheds of the lower Sutlej River basin. *Times of India*. https://timesofindia.indiatimes.com/blogs/darksides/spotlight-on-erosion-prioritizing-the-watersheds-of-the-lower-sutlej-river-basin/

Florsheim, J. L., Mount, J. F., & Chin, A. (2008). Bank erosion as a desirable attribute of rivers. *BioScience, 58*(6), 519–529. https://doi.org/10.1641/B580608

Garde, R. J. (2006). *River morphology*. New Age International Publishers.

Goswami, J., Velliyangiri, S., Sarma, K. K., & Aggarwal, S. (2024). Assessment of agricultural land area lost due to Brahmaputra riverbank line erosion in Assam. *Current Science, 127*(1), Article 0020.

Gupta, P. (2025). The ultimate 2024 guide to saving the Lower Sutlej River Basin. *Times of India*. https://timesofindia.indiatimes.com/blogs/darksides/the-ultimate-2024-guide-to-saving-the-lower-sutlej-river-basin/

Horton, R. E. (1933). The relation of hydrology to the botanical sciences. *Eos, Transactions American Geophysical Union, 14*(1), 23–25.

Horton, R. E. (1945). Erosional development of streams and their drainage basins: Hydrophysical approach to quantitative morphology. *Geological Society of America Bulletin, 56*(3), 275–370.

Kaur, S., Gani, A., Pathak, S., & Tiwari, R. K. (2025). Morphometric analysis of sand mining-induced changes in the Sutlej River. *Environmental Geochemistry and Health, 47*(11), 459.

Langat, P. K., Kumar, L., & Koech, R. (2019). Monitoring river channel dynamics using remote sensing and GIS techniques. *Geomorphology, 325*, 92–102. https://doi.org/10.1016/j.geomorph.2018.10.007

Philip, P., Baby, N., Ashwathi, V. K., & Thottungal, D. (2014). Morphometric analysis of Kallada and Pallikkal river basins of Kerala using GIS. *International Journal of Innovative Research in Science, Engineering and Technology, 3*(4), 11530–11537.

Rozo, M. G., Nogueira, A. C., & Castro, C. S. (2014). Remote sensing-based analysis of the planform changes in the upper Amazon River over the period 1986–2006. *Journal of South American Earth Sciences, 51*, 28–44. https://doi.org/10.1016/j.jsames.2013.12.004

Saeedrashed, Y., & Guven, A. (2012). Estimation of geomorphological parameters of lower Zab river-basin by using GIS-based remotely sensed image. *Water Resources Management, 27*(1), 209–219. https://doi.org/10.1007/s11269-012-0179-x

Sarkar, A., Garg, R. D., & Sharma, N. (2012). RS-GIS based assessment of river dynamics of Brahmaputra River in India. *Journal of Water Resource and Protection, 4*(2), 63–72. https://doi.org/10.4236/jwarp.2012.42008

Sarma, J., & Phukan, M. (2004). Origin and some geomorphological changes of Majuli island of the Brahmaputra River in Assam, India. *Geomorphology, 60*(1–2), 1–19. https://doi.org/10.1016/j.geomorph.2003.07.013

Setia, R., Dhaliwal, S. S., Kumar, V., Singh, R., Kukal, S. S., & Pateriya, B. (2020). Impact assessment of metal contamination in surface water of Sutlej River (India) on human health risks. *Environmental Pollution, 265*, 114907. https://doi.org/10.1016/j.envpol.2020.114907

Singh, A., Sharma, C. S., Jeyaseelan, A. T., & Chowdary, V. M. (2015). Spatio–temporal analysis of groundwater resources in Jalandhar district of Punjab state, India. *Sustainable Water Resources Management, 1*(3), 293–304. https://doi.org/10.1007/s40899-015-0027-1

Stephens, C. M., Lall, U., Johnson, F. M., & Marshall, L. A. (2021). Landscape changes and their hydrologic effects: Interactions and feedbacks across scales. *Earth-Science Reviews, 212*, 103466. https://doi.org/10.1016/j.earscirev.2020.103466

Strahler, A. N. (1957). Quantitative analysis of watershed geomorphology. *Transactions of the American Geophysical Union, 38*(6), 913–920.

Strahler, A. N. (1964). Quantitative geomorphology of drainage basins and channel networks. In V. T. Chow (Ed.), *Handbook of applied hydrology* (pp. 439–476). McGraw-Hill.

Trent, R. E., & Brown, S. A. (1984). An overview of factors affecting river stability. *Transportation Research Record, 950*(2), 156–163.

Sediment Load Estimation Using Geostatistical Method

Om Prakash Biswal, Anindita Swain, and Laxmipriya Mohanty

Abstract Ungaged (data-scarce) basin lack access to daily sediment data and to estimate it, a popular technique is adopted in which the characteristics of a reference basin's sediment flow time series are transferred to the data-scarce one. In conventional methods, the data-scarce basin's parameters are typically determined by using a reference station that is closest to that basin. The close neighbor station may not always be the one with the highest correlation, it is established in the current study. To find the reference basin which is the highest correlated, an approach known as the Map Correlation Method (MCM) is used. The effectiveness of current approach is evaluated using daily sediment data collected over a five-year period at 17 sediment gaging sites in Krishna and Godavari watersheds. Pearson's *r* correlation coefficient's geographic structure is modelled using sample and model variograms. Selection of basin with optimum correlation value is taken as reference station in ungagged locations. Observed and estimated correlations resulting from the cross-validation of the variograms are in line with each other for most of the stations.

Keywords Geospatial technique · Kriging · Map correlation method · Ungaged basin · Variogram

1 Introduction

Rainfall is responsible for soil erosion whereas runoff helps in transport of eroded soil as sediment (Heng & Suetsugi, 2013). Out of the total sediment, maximum contribution is from suspended load which is about 90% (Babi, n.d.) whereas remaining is from bed load (Walling & Fang, 2003). The effect of sediment load can be seen

O. P. Biswal · L. Mohanty (✉)
Department of Civil Engineering, Veer Surendra Sai University of Technology, Burla, Sambalpur, Odisha, India
e-mail: lmohanty_ce@vssut.ac.in

A. Swain
Department of Civil Engineering, GIET, BBSR, Khurda, Odisha, India

S. Pathak and M. Pandey (eds.), *Geospatial Technology and Advanced Instrumentation in Addressing Water Resources Planning and Management*, Water Informatics for Water Resource Management, https://doi.org/10.1007/978-981-92-1602-4_5

in many areas such as: capacity of hydraulic structures, problems in aquatic habitat, river geometry modification, and water quality degradation (Mohanty & Biswal, 2022; Mohanty et al., 2022b). Hence, proper estimation of sediment load is quite necessary for both field engineers and research persons. However, most of the basins are gaged according to discharge but ungaged for sediment globally (Mohanty et al., 2023, 2025). Because it is not always feasible to conduct experiments due to several reasons such as inaccessible site locations and complex experimental procedures (Bhuyan et al., 2024).

The other way out for sediment load estimation is through several modeling approaches like physical, conceptual, or data-driven methods. Physical models are quite robust because they are based on underlying physics behind the phenomena but it has large input data and huge computational resource. Similarly, data-driven black box techniques give some output without bothering about the background physics. Conceptual models serve a midway among all but only when the site is gaged. Moreover, for ungaged locations, the geospatial techniques are quite suitable for parameter estimation (Mohanty et al., 2022a, 2022b).

In 1979, Hirsch (Hirsch, 1979) selected two reservoirs from the Roanoke River at Roanoke gage where 49-year of data of stream flow were available. But for the sake of this study, it was assumed that the considered dataset did not exist or only limited data available. Based on this assumption the stream flow records for these sites were tried to be reconstructed. The reconstruction of stream flow records is based on the transfer of data by cross-correlation of flow records from nearby stream gages, and estimates of monthly mean and standard deviations derived from regional stream flow-basin characteristics models. The study concluded useful estimates of reservoir yield can be made by the transfer of data from closest basins when there is a limitation of site data. An ungaged catchment's daily stream flow can be calculated by using a reference stream gage (Sivapalan, 2003) using the Drainage Area Ratio (DAR) method. But however, to do this efficient reconstruction rigorous development and testing of this technique are required (Hirsch, 1979). Similarly, the U.S. Geological Survey launched a study project, in 1984. The annual SS loads of a few selected Colorado streams were calculated with the help of the available sediment data (Ryan et al., 1986) in the first stage of this study program. The program's second phase entails the creation of a method for calculating the average annual SS loads for streams where there is absence of the sediment or stream flow data. Numerous studies have been done to determine the daily stream flow in ungaged basin using a donor gage (Sivapalan, 2003). The general, and arguably first, approach for estimating the daily stream flow of an ungaged basin using a donor basin is the DAR method. Furthermore, a modified DAR that standardizes monthly stream flows has been presented by Hirsch (1979). The neighboring stream gage is often considered as the reference station (Mohamoud, 2008; Smakhtin, 1999; Smakhtin et al., 1997).

There is a lack of studies on predicting daily sediment discharge (SD) in ungaged catchments. One approach to estimate the mean annual suspended sediment load (SSL) is by averaging the measured daily suspended sediment discharge (SSD). Another method involves using a sediment transport curve (STC) derived from periodic measurements of SSD to calculate the annual mean SSL (Mohanty et al., 2022a;

Ryan et al., 1986). Multiple linear regression models are used to relate the mean yearly sediment load (SL) of streams with large sediment dataset to the basin characteristics, allowing for the determination of mean yearly SL for ungaged rivers in the same region. However, obtaining a substantial dataset for sediment prediction models can be challenging, which is a must for the prediction models of sediment or erosion, as sediment data collection is expensive and requires consistent monitoring stations (Walling & Fang, 2003). Many sites face conditions that make estimating sediment in stream discharge difficult, and data obtained from different agencies may show discontinuity over space and time. This has led to the concept of ungaged stations, where stream flow data is unavailable, scarce, discontinuous, or absent. Stream flow gaging sites often exclude recorded SS data due to factors such as significant stream flow regulation, unavailability of long-term records, high variability in discharge and SS concentrations, or questions about data accuracy (Ryan et al., 1986). The absence of continuous daily SS data at many gaging locations hampers time series analysis. To maintain consistency in sediment yield (SY) data in such cases, literature recommends reconstructing stream flow records based on data transfer from nearby known stream gages. Various models, including regional and empirical models, can be used to calculate stream flow (Arsenault & Brissette, 2014), that correlate to DAR (Msadala & Basson, 2017). However, it is found that the SY of an unmeasured catchment may not be accurately represented by the nearest sediment gage station. The Map Correlation Method (Mohanty et al., 2022a) is a geostatistical technique that utilizes the Pearson correlation coefficient and Euclidean distance between two sediment gage stations to evaluate their association (Rouhani et al., 1996). This method is tested on a gaged catchment assuming its records to be limited or not existing to assess its effectiveness.

2 Study Area and Data Collection

The amount of SS transported annually in a river is known as the "Annual SS load" and is determined by summing up the daily SS flows for that specific year. The daily SS discharge can be calculated using a rating curve (i.e., a curve between the SS concentration and water discharge) or by quantifying the SS present in the daily water flow (Porterfield, 1972). However, recording daily discharge of SS is not common in Indian catchments due to the high cost involved. Despite this, many small catchments in India do maintain data on daily and yearly stream flows. The study focuses on 17 catchments in the Krishna and Godavari Rivers in South India. Out of these, 10 catchments are from the Krishna River, and 7 are from the Godavari River. Five years of daily sediment and flow record, from 1995 to 2000, are used in this work. The lack of more gages is due to the difficulty in fetching observed data. Fortunately, the required data on daily stream flow and sediment concentration are available for these 17 catchments. For more detailed information on the 17 catchments in the Krishna and Godavari watersheds, Table 1 provides an in-depth list.

Table 1 Details of catchment in Krishna and Godavari basin

Sl. No.	River name	Basin name	Area (km^2)	Lat.	Long.
1	Krishna	Bawapuram	67,180	15.88	77.95
2	Krishna	Haralahalli	14,582	14.83	75.67
3	Krishna	Honnali	7075	14.23	75.65
4	Krishna	Karad	5462	17.29	74.19
5	Krishna	Keesara	9854	16.71	80.32
6	Krishna	Malkhed	7650	17.2	77.15
7	Krishna	Pondugala	221,220	16.68	79.66
8	Krishna	Shimoga	2831	13.93	75.57
9	Krishna	Wadenapally	235,544	16.79	80.07
10	Krishna	Yadgir	69,863	16.73	77.12
11	Godavari	Degloor	2869.21	18.53	77.58
12	Godavari	Jagdalpur	7380	19.11	82.02
13	Godavari	Konta	19,550	17.8	81.38
14	Godavari	Nowrangpur	9933	19.2	82.51
15	Godavari	Perur	130,407	18.55	80.366
16	Godavari	Polavaram	307,800	17.24	81.65
17	Godavari	Saigaon	13,558.426	18.05	77.05

3 Methodology

3.1 Map Correlation Method (MCM)

The MCM suggests a donor station for a data-scarce basin based on the highest correlation, regardless of proximity. It assumes (i) hydrologic similarity between daily SY and correlation coefficients, and (ii) isotropy of correlation coefficients in the study area. Spatial covariance of each variogram is calculated using leave-one-out cross-validation, and a variogram is used to replicate the spatial structure of correlation coefficients with each stream gage. This method requires fewer stream gages and does not assume homogeneous areas. The MCM method employs ordinary kriging to create a continuous map showing correlation between the pseudo-ungaged basin and the remaining gaged basins. Pseudo-ungaged basins actually gauged but are treated as ungaged for cross-checking. Correlation maps are created for each gaged catchment with the pseudo-ungaged catchment. The reference catchment is selected based on the highest correlated map with the pseudo-ungaged catchment.

4 Result Analysis

4.1 Relation Between Distance and Correlation of SY

The main objective of this paper is to understand the correlation between log-transformed daily SY values at the study gages and the distance between the gages. Based on the closest gage basin, the reference stream gage is suggested, which is when the Pearson's r and distance are exactly correlated. Here, generally the nearest basin shows the highest correlation with the ungaged basin. The Euclidean distance is considered as the representative of distance between donor and ungaged basin, and log-transformed SY data between the ungaged and highest correlated catchment is used to calculate Pearson's correlation coefficient (r). Table 2 provides a list of the most correlated and the nearest reference stream gages. Daily SY data from 17-gage stations in Krishna and Godavari watersheds during the five consecutive years of time period (1995–2000) were considered. Although in this study Pearson's r is recommended, Kendall's τ can be helpful in cases when monotonic relations exist (Helsel et al., 2020). Figure 1 shows the relationship between distance and correlation for each of the four selected research gages. Pearson's correlation coefficient (r) was calculated using logarithms for each set of five-year stream-gage records from January 1, 1995, to December 31, 2000. Interestingly, the closest catchment is not always the highest correlated one, indicating that distance cannot always be substituted for correlation. The goodness of fit is measured using the NSE value (Nash & Sutcliffe, 1970). The NSE values of the closest gage and the highest correlated stream gage are compared in Table 3 to choose the reference stream gage. Figure 2 shows the arithmetic data sets, while Fig. 3 shows the log-transformed data sets, illustrating the range of NSE values for the most correlated and closest catchments. The figure suggests that, except for a few gage stations, the NSE values are significantly improved when comparing the highest correlated stations to the closest ones.

In the study's examined gage basins, poor NSE values were observed in basins lacking correlated gages. To assess uncertainty in ungaged stream gages, the relationship between Pearson's r and NSE was used. NSE values were calculated using log transformed values from the nearest and most correlated sites. These values were then graphed with distance and Pearson's r (Fig. 3). If NSE decreases with distance, the closest gaged station should be considered as the reference stream gage. However, Fig. 4a does not show this pattern. In contrast, Fig. 4b reveals a discernible pattern in Pearson's r plot, indicating that the donor gage should be the one with the highest correlation, rather than the closest one, for a more accurate forecast of SY with minimal scatter.

Table 2 List of nearest and most correlated reference stream-gage

Sl. No.	Basin	Nearest station	Distance (km) of nearest station from the study basin	Most correlated station	Pearson's *r*
1	Krishna-Bawapuram	Yadgir	117.5	Perur	0.42
2	Haralahalli	Honnali	59.2	Perur	0.58
3	Honnali	Shimoga	29.6	Karad	0.12
4	Karad	Saigaon	229.6	Shimoga	0.13
5	Keesara	Wadenapally	25.9	Perur	0.48
6	Malkhed	Yadgir	47.5	Shimoga	0.13
7	Pondugala	Wadenapally	43.1	Shimoga	0.32
8	Shimoga	Haralahalli	90	Pondugala	0.32
9	Wadenapally	Keesara	25.9	Haralahalli	0.15
10	Yadgir	Malkhed	47.5	Konta	0.19
11	Godavari-Degloor	Saigaon	72.2	Polavaram	0.12
12	Jagdalpur	Nowrangpur	50.2	Konta	0.16
13	Konta	Polavaram	62.1	Yadgir	0.19
14	Nowrangpur	Jagdalpur	50.2	Polavaram	0.28
15	Perur	Konta	126.2	Polavaram	0.91
16	Polavaram	Konta	62.1	Perur	0.91
17	Saigaon	Degloor	72.2	Perur	0.11

4.2 Outputs of Map Correlations for Study Stream Gages

For an accurate calculation of correlation, each stream gage's record must coincide with others for a long period of time. A reference streamgage Perur's correlation with 16 other streamgages is geographically illustrated in Fig. 5. Figure 6b displays the derived correlation values for all 17 stream gages while taking into account Fig. 6a, which depicts a spherical-type semi-variogram model.

Semivariance is determined for each group of stream gages in the watershed as the difference in squared *r* values and the related Euclidean distance (Gorsich & Genton, 2000). A continuous function is described by the separation distance and semivariance, which are plotted to create a variogram cloud (Hohn, 1991). The term "cloud" is typically used to describe this figure, which consists of (*n* 2) points and has no perceptible pattern, where n is the number of gage stations, which in this instance is 17.

A sample variogram can be created by binning scattered points within a specific separation distance. For each bin, the mean value represents the semivariance. Model variograms are created using sample variograms that were created during the binning

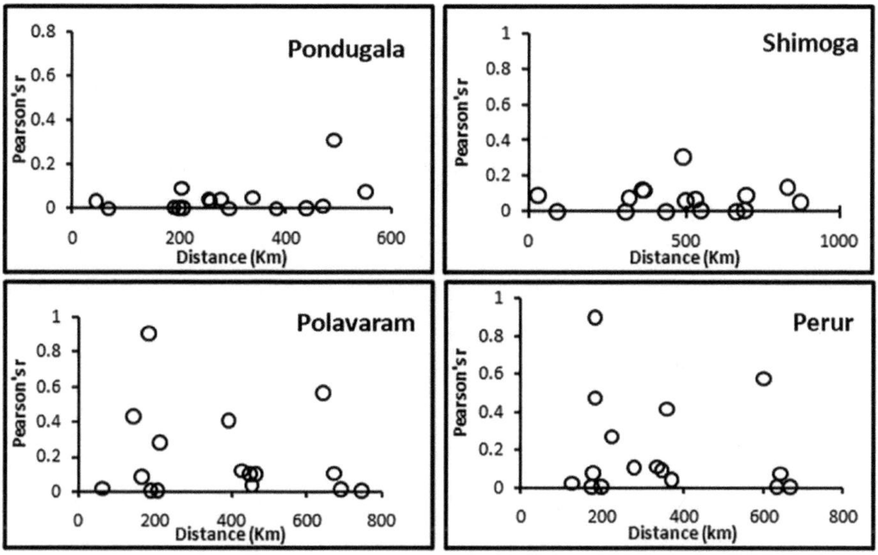

Fig. 1 Relation between distance and Pearson's *r* for 4 sample stream gages of the Krishna-Godavari basin

process. Bin lengths were manually adjusted through trial and error. For each of the 17 stream gages, the bin length was used to match the variogram model. Based on the number of points drawn to estimate each location on the sample variogram, the weights used in the weighted least squares approach are determined, which is then used to fit the variogram model (Ribeiro & Diggle, 2001). The spherical type has been chosen from among the various variogram models because it is simple to formulate and visualize for sample variograms at every stream gage. The spherical variogram equations (Ribeiro & Diggle, 2001) can be used to determine the variogram model, separation distance, and range. The covariance function was then calculated (Ribeiro & Diggle, 2001) using conventional geostatistical kriging (Hohn, 1991). The spatial covariance structure was discovered using the geo-R program suite. Each sample stream gage was subjected to a leave-one-out cross-validation trial for this. Based on cross-validation and variogram model results, the *r* values for the considered basins follow a clear geographic pattern.

16 *r* values were methodically cleared from the experimental variogram for each of the 17 variogram models given in Table 4, and then the variogram model parameters were once again estimated. The *r*-value for each basin that had been deleted by a model variogram, was revaluated by comparing it to the actual value. To determine the NSE value, which is used to rate the goodness of fit for each of the 17 variogram models, 16 amounts of predicted and observed *r* values were used. The geographic relationship between the cross-validation findings and the *r*-values for a specific stream gage can be examined using the variogram models. Geostatistics can be used

Table 3 Comparison of NSE values by adopting reference station the nearest stream gage, and the stream gage having the most correlated sediment yield

River name	Study stream gage	NSE value computed from arithmetic-transformed observed and estimated values for the reference stream gage		NSE value computed from log-transformed observed and estimated values for the reference stream gage	
		Nearest stream gage	Most correlated stream gage	Nearest stream gage	Most correlated stream gage
Krishna	Bawapuram	− 0.77	− 0.71	0.12	0.06
	Haralahalli	− 0.45	− 0.74	0.81	0.79
	Honnali	− 0.76	− 0.62	− 0.56	0.74
	Karad	− 0.09	0.46	0.5	0.25
	Keesara	− 0.18	− 0.27	0.93	0.69
	Malkhed	− 0.43	0.81	− 0.22	0.91
	Pondugala	0.5	0.62	0.85	0.62
	Shimoga	0.93	0.59	0.77	0.82
	Wadenapally	0.8	− 0.55	0.12	0.59
Godavari	Yadgir	0.04	0.5	0.9	0.58
	Degloor	0.1	0.43	0.82	− 0.2
	Jagdalpur	0.83	− 0.88	− 0.14	0.86
	Konta	− 0.22	− 0.52	− 0.74	0.9
	Nowrangpur	0.36	− 3.6	− 0.39	0.15
	Perur	0.04	0.46	− 0.24	0.81
	Polavaram	0.04	0.63	0.64	0.72
	Saigaon	0.07	− 5.9	0.88	0.61

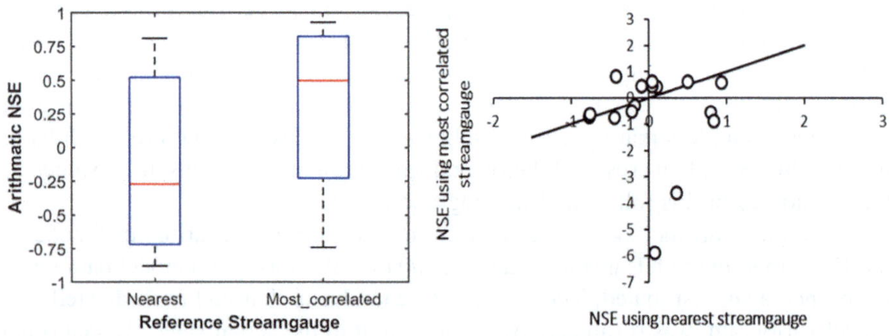

Fig. 2 NSE values for stream gage stations using arithmetic-transform data

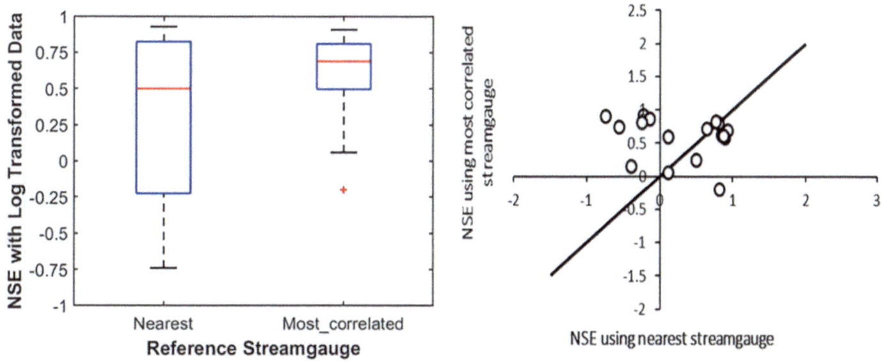

Fig. 3 NSE values for stream gage stations using log-transform data

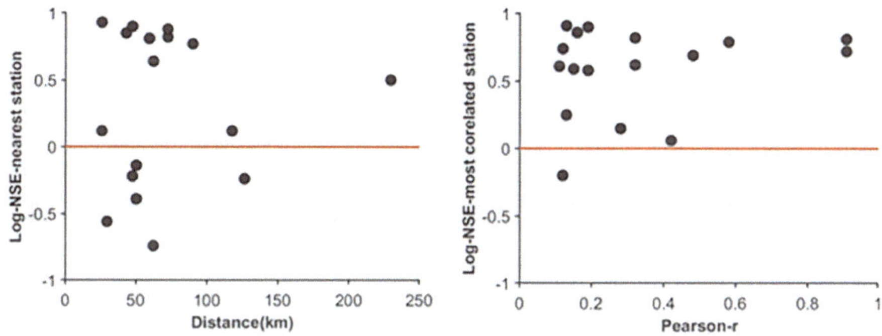

Fig. 4 Relation between NSE value with distance, and Pearson's *r*-correlation coefficient

to determine the reliability of the majority of stream gages, but some may have a stronger spatial structure than others.

The model variogram provides the covariance characterizing the geographic relationship between *r* values and separation distance as well as its parameters. The inputs are used to perform ordinary kriging to calculate the *r* value between each ungaged catchment and the remaining stream gages. Using the calculated covariance function and parameters, a 17×17 array of the covariance matrix may be created, showing the separation distances between each pair of stream gages. Additionally, the distance between the 17 stream gages and the data-scarce catchment is calculated. The covariance function is then applied over this distance to create a 17×1 matrix. By multiplying the 17×17 matrices acquired above with the transposed 17×1 matrix, a set of 17 weights are determined using ordinary kriging. Finally, the correlation between the study stream gage and the data-scarce catchment is calculated by multiplication of these weights with the corresponding *r* values. Hence, ordinary kriging can be used to produce a map correlation between any basin and the entire

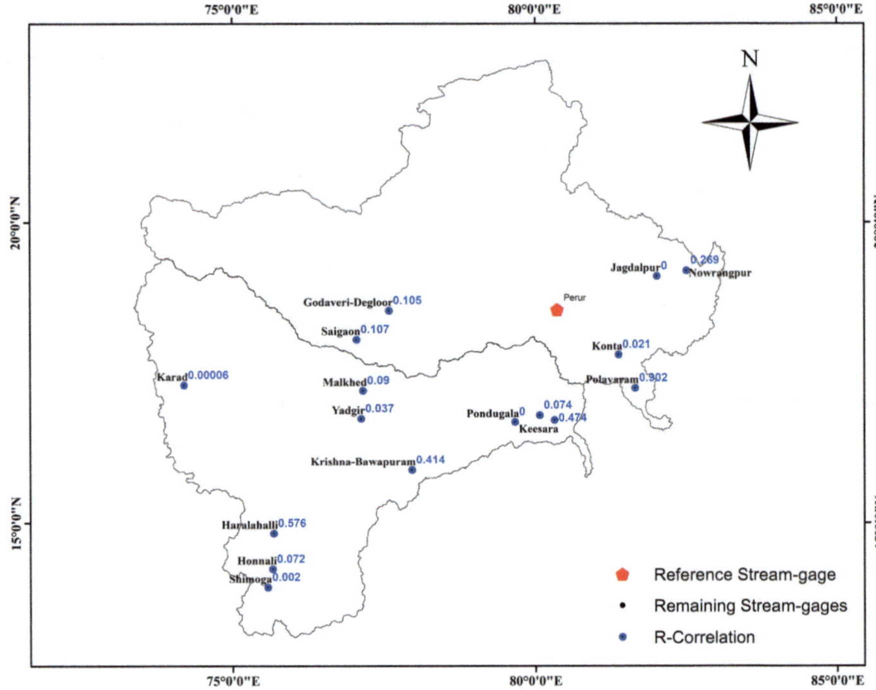

Fig. 5 Pearson's r correlation coefficient values computed between the logarithms of the daily sediment yield time series at Perur streamgage and 16 other streamgages in Krishna and Godavari basin

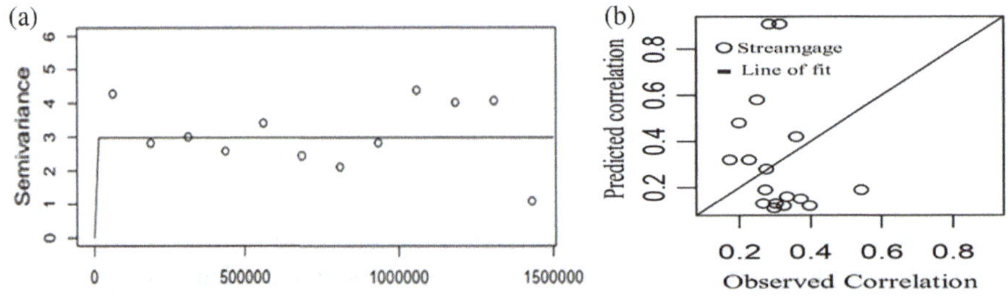

Fig. 6 **a** Sample and model variograms used to model the spatial structure of Pearson's r correlation coefficient with the Perur station. **b** Observed and estimated correlations resulting from a leave one-out cross-validation of the respective variograms

study area. For datasets containing zero values, Pearson's r should be calculated with extra caution (Koutsoyiannis et al., 2008).

Table 4 Spherical variogram model parameters calculated from the Pearson's *r*-correlation coefficient

River name	Study stream gage	Variance parameter	Range parameter
Krishna	Bawapuram	0.0038	132,292.3
	Haralahalli	0.0033	192,391.2
	Honnali	0.0035	152,123.1
	Karad	0.0028	141,123.1
	Keesara	0.0031	212,282.1
	Malkhed	0.0058	112,311.1
	Pondugala	0.0048	101,101.2
	Shimoga	0.004	121,131.1
	Wadenapally	0.0032	112,141.2
	Yadgir	0.0038	142,210.1
Godaveri	Degloor	0.0025	151,101.2
	Jagdalpur	0.0031	201,321.1
	Konta	0.0082	283,124.2
	Nowrangpur	0.0052	142,213.2
	Perur	0.009	182,936.3
	Polavaram	0.0056	123,210.3
	Saigaon	0.0025	145,362.1

4.3 Use of MCM in Ungaged Basins

The highest correlated stream gage found in MCM is considered as the reference stream gage. By calculating the correlation between the study gaged basin and the reference gaged basin using daily time series SY data, the highest correlated gage basin for each pseudo-ungaged basin was identified in the previous section. However, with the actual ungauged data, the highest correlated catchment is found using the MCM technique. The highest correlated basins for Pseudo ungaged basins are estimated in a similar manner here using the Map correlation approach, and then it is compared to see if they match the highest correlated basins determined manually.

For most of the study catchments, the MCM method's highest correlated catchment matches the manually calculated one. Despite the fact that the computed NSE values and the previously acquired NSE values diverge, this variation has no larger impact on the choice of the donor basin. Figure 7 depicts the Kriging relationship between a hypothetical reference stream gage and an ungaged basin's daily SY. Figure 7 displays a kriged map of the Pearson's *r* correlation coefficient for 16 basins from Perur, an important reference stream gage.

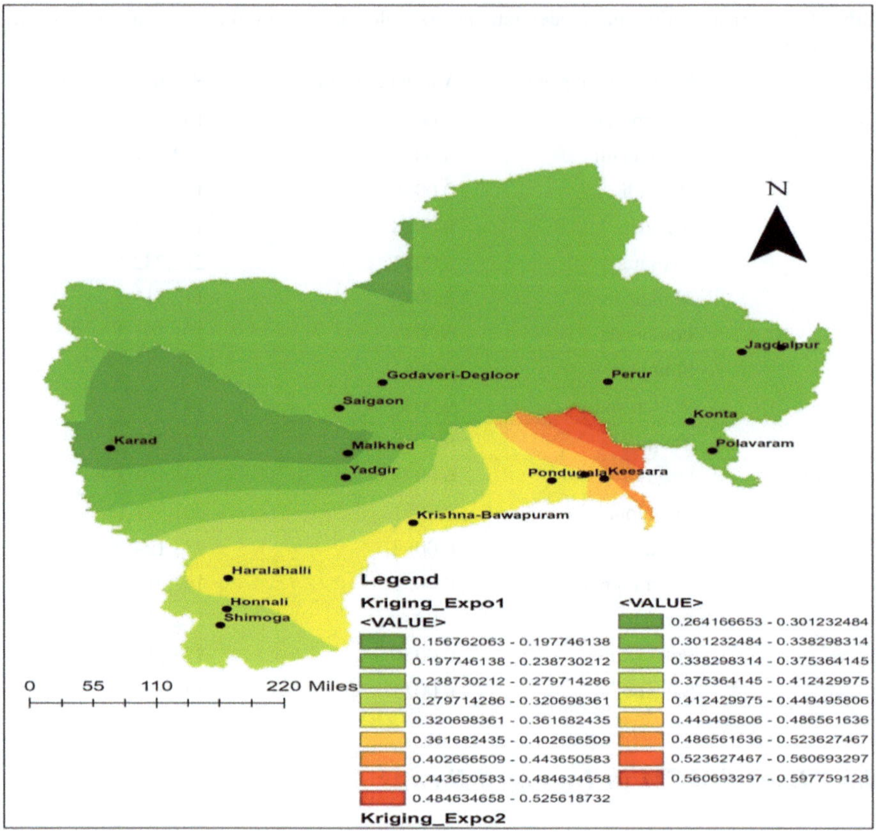

Fig. 7 Map showing the spatial distribution of Pearson's *r* correlation coefficient values estimated between the daily sediment yield time series at Perur and all other streamgages in Krishna-Godavari basin

4.4 Discussion on NSE

The NSE value was calculated for goodness of fit test. At first this calculation was used to find the observed reference stream gage in between nearest and most correlated stream gages. Then this calculation was used to cross verify that whether the predicted highest correlated basin coincides with the observed or not. Table 5 displays the reference stream gage that was produced using the map correlation technique. The gap in the *r* values for the stream gage selected using MCM and the basin with the highest correlation are also tabulated. In four instances, the differences can be seen to be zero, and for a few other gage sites, there is a slight difference that can be overlooked.

Table 5 Reference stream gages identified from MCM method for estimation of discharge at different ungaged stations

River name	Study stream gage	MCMM most correlated stream gage	The difference of r between the most correlated stream gage and stream gage selected by map correlation method
Krishna	Bawapuram	Perur	0.001
	Haralahalli	Perur	0.03
	Honnali	Karad	0.002
	Karad	Shimoga	0.01
	Keesara	Perur	0.2
	Malkhed	Shimoga	0
	Pondugala	Shimoga	0
	Shimoga	Pondugala	0.001
	Wedenapally	Haralahalli	0
	Yadgir	Konta	0.002
Godavari	Degloor	Polavaram	0.006
	Jagdalpur	Konta	0.05
	Konta	Yadgir	0.1
	Nowrangpur	Polavaram	0.01
	Perur	Polavaram	0.001
	Polavaram	Perur	0
	Saigaon	Perur	0.03

5 Conclusion

To forecast sediment output in an ungaged catchment, a number of numerical or conceptual approaches are now available. Selecting a reference stream gage from the neighborhood gage stations in a way that allows for data transfer with the least amount of error is a challenging job, though. Our research led us to the conclusion that picking a reference stream gage based on distance is not always a good idea. The DAR approach produced the same results as well. However, using the catchment with the greatest correlation as a reference can greatly improve the predicted values. The Map Correlation Method's results have been incorporated into this theory as reference stream gages.

Geostatistical modeling was successful due to the research area's well-built spatial relationships between correlations. Spatial models were used to evaluate the correlation between ungaged and a group of gaged basins in order to identify the reference stream gage. The daily SY time series could be enhanced using the MCM method. The Map Correlation technique can help to classify catchments, calibrate hydrologic models in ungaged basins, and identify uniqueness in a network of stream gage.

References

Arsenault, R., & Brissette, F. P. (2014). Continuous streamflow prediction in ungauged basins: The effects of equifinality and parameter set selection on uncertainty in regionalization approaches. *Water Resources Research, 50*, 6135–6153. https://doi.org/10.1002/2013WR014898

Babi, Z. (n.d.). The relationship between suspended and bed load transport in river channels.

Bhuyan, P. K., Meher, J., & Mohanty, L. (2024). Soil erosion estimation using RUSLE: A case study of Bamnidhi sub basin. *ISH Journal of Hydraulic Engineering, 30*(2), 273–280. https://doi.org/10.1080/09715010.2024.2301721

Gorsich, D. J., & Genton, M.G. (2000). *Variogram model selection via nonparametric derivative estimation, mathematical geology.*

Helsel, D. R., Hirsch, R. M., Ryberg, K. R., Archfield, S. A., & Gilroy, E. J. (2020). Statistical methods in water resources. https://doi.org/10.3133/tm4A3

Heng, S., & Suetsugi, T. (2013). Using artificial neural network to estimate sediment load in ungauged catchments of the Tonle Sap River Basin, Cambodia. *JWARP, 05*, 111–123. https://doi.org/10.4236/jwarp.2013.52013

Hirsch, R. M. (1979). An evaluation of some record reconstruction techniques. *Water Resources Research, 15*, 1781–1790. https://doi.org/10.1029/WR015i006p01781

Hohn, M. E. (1991). An Introduction to Applied Geostatistics: by Edward H. Isaaks, R. Mohan Srivastava, 1989, Oxford University Press, New York, 561 p., ISBN: 0-19-505012-6, ISBN 0-19-505013-4 (paperback), $55.00 cloth, $35.00 paper (US). *Computers & Geosciences, 17*, 471–473. https://doi.org/10.1016/0098-3004(91)90055-I

Koutsoyiannis, D., Yao, H., & Georgakakos, A. (2008). Medium-range flow prediction for the Nile: A comparison of stochastic and deterministic methods / Prévision du débit du Nil à moyen terme: Une comparaison de méthodes stochastiques et déterministes. *Hydrological Sciences Journal, 53*, 142–164. https://doi.org/10.1623/hysj.53.1.142

Mohamoud, Y. M. (2008). Prediction of daily flow duration curves and streamflow for ungauged catchments using regional flow duration curves. *Hydrological Sciences Journal, 53*, 706–724. https://doi.org/10.1623/hysj.53.4.706

Mohanty, L., & Biswal, B. (2022). Event scale analysis of sediment Concentration-River discharge relationship. *Materials Today Proceedings, 62*(12), 6379–6384. https://doi.org/10.1016/j.matpr.2022.03.383

Mohanty, L., Istalkar, P., & Biswal, B. (2023). Dynamic aspects of suspended-sediment-concentration recession curves. *Journal of Hydrology, 617* (Part C), 129107. ISSN: 0022-1694. https://doi.org/10.1016/j.jhydrol.2023.129107

Mohanty, L., Swain, A., & Nanda, B. (2022a). Estimation of parameters in ungaged catchment using map-correlation method: A case study on Krishna-Godavari basin. In S. Kanga, S. K. Singh, G. Meraj, & M. Farooq (Eds.), *Geospatial modeling for environmental management*. Case studies from South Asia (pp. 19–36). ISBN: 9781003147107. https://doi.org/10.1201/9781003147107

Mohanty, L., Swain, A., Patnaik, M., & Biswal, B. (2022b). Sediment-yield prediction using map correlation method. *Materials Today Proceedings, 62*(12), 6391–6400. ISSN: 2214-7853. https://doi.org/10.1016/j.matpr.2022.03.504

Mohanty, L., Qureshi, A., & Biswal, B. (2025). Finding a parsimonious suspended sediment transport model structure. *Journal of Hydrology, 657*, Article 133143. https://doi.org/10.1016/j.jhydrol.2025.133143

Msadala, V., & Basson, G. (2017). Revised regional sediment yield prediction methodology for ungauged catchments in South Africa. *Journal of the South African Institution of Civil Engineering, 59*, 28–36. https://doi.org/10.17159/2309-8775/2017/v59n2a4

Nash, J. E., & Sutcliffe, J. V. (1970). River flow forecasting through conceptual models. Part I—A discussion of principles. *Journal of Hydrology, 10*, 282–290. https://doi.org/10.1016/0022-1694(70)90255-6

Porterfield, G. (1972). Computation of fluvial-sediment discharge. https://doi.org/10.3133/twri03C3

Ribeiro, P. J., Jr., & Diggle, P. (2001). *geoR: A package for geostatistical analysis* (Vol. 1).

Rouhani, S., Mohan Srivastava, R., Desbarats, A. J., Cromer, M. V., & Ivan Johnson, A. (1996). *Geostatistics for environmental and geotechnical applications*. ASTM STP 1283. ASTM. https://archives.colostate.edu/repositories/2/archival_objects/87398. Accessed April 3, 2023.

Ryan, S. E., Shobe, C. M., Rathburn, S. L., & Dixon, M. K. (1986). *Sediment-data sources and estimated annual suspended-sediment loads of rivers and streams in Colorado*. https://doi.org/10.3133/wri864344

Sivapalan, M. (2003). Prediction in ungauged basins: A grand challenge for theoretical hydrology. *Hydrological Processes, 17*, 3163–3170. https://doi.org/10.1002/hyp.5155

Smakhtin, V. Y. (1999). Generation of natural daily flow time-series in regulated rivers using a non-linear spatial interpolation technique. *Regulated Rivers: Research & Management, 15*, 311–323. https://doi.org/10.1002/(SICI)1099-1646(199907/08)15:4%3c311::AID-RRR544%3e3.0.CO;2-W

Smakhtin, V. Y., Hughes, D. A., & Creuse-Naudin, E. (1997). Regionalization of daily flow characteristics in part of the Eastern Cape, South Africa. *Hydrological Sciences Journal, 42*, 919–936. https://doi.org/10.1080/02626669709492088

Walling, D. E., & Fang, D. (2003). Recent trends in the suspended sediment loads of the world's rivers. *Global and Planetary Change, 39*, 111–126. https://doi.org/10.1016/S0921-8181(03)00020-1

Implementing LID Techniques for Runoff Reduction Through the Curve Number Method Considering Slope Variation in Uttarakhand

Shray Pathak and Cheshta Bhardwaj

Abstract Uttarakhand, a Himalayan state in India, is significantly affected by the flooding in urban regions due to high intensity rainfall, varied land use/ land cover (LULC) and steep topography. This chapter calculates direct runoff estimation by Soil Conservation Service Curve Number (SCS-CN) technique combining LULC, hydrological soil groups (HSG), rainfall and slope data, in ArcMap. Curve Number method formula was used to introduce slope effect on Curve Number (CN) values. In order to reduce runoffs, the Low Impact Development Techniques have been implemented through installing permeable and rain gardens by adjusting CN to capture greater infiltration. The runoff was recalculated after the implementation of LIDs. The results emphasize the effectiveness of integrating slope-based CN modifications and LID practices to reduce surface runoff. This method is used to illustrate the way that sustainable land use planning and hydrological modelling can be combined for better flood risk management in mountainous areas. The methodology offers a replicable model for other regions that are experiencing similar stresses of rapid urbanization and climate change-induced hydrological extremes.

Keywords Land Use Land Cover (LULC) · Hydrologic soil group (HSG) · Curve number (CN) · Runoff estimation · Low impact development (LID) techniques

1 Introduction

Flood risks in the Himalayan state of Uttarakhand that has been shaped by steep slope, heterogeneous LULC, and erratic rainfall during monsoon seasons have been aggravated by urbanization and climate change (Kansal & Singh, 2022). One of the

S. Pathak
Department of Civil Engineering, Indian Institute of Technology, Ropar, India
e-mail: shray.pathak@iitrpr.ac.in

C. Bhardwaj (✉)
Department of Geography, Panjab University, Chandigarh, India
e-mail: cheshta.h@gmail.com

S. Pathak and M. Pandey (eds.), *Geospatial Technology and Advanced Instrumentation in Addressing Water Resources Planning and Management*, Water Informatics for Water Resource Management, https://doi.org/10.1007/978-981-92-1602-4_6

major causes of flooding in an urban area is direct runoff and some of the factors that influence them include impervious land coverage, the infiltration ability of soil, rainfall intensity and topography. The Natural Resources Conservation Service Curve Number (NRCS-CN), formerly known as soil conservation services Curve Number, is the most widely used method to estimate direct runoff from rainfall, due to its simplicity and the use of the single CN parameter (Garg et al., 2013). The first step in using these methods is to determine the Curve Number (CN), which is a value between 1 and 100, for the modeled watershed. The CN modulates the conversion of rainfall volume into runoff volume, with larger CNs estimating greater runoff. Forested watersheds and other natural systems that allow considerable infiltration may have CNs as low as the 40 s and 50 s, whereas highly urbanized systems with impervious surfaces may have CNs in the 80 s or 90 s. Determination of the CN is often the first of several calculations required in rainfall-runoff modeling, with subsequent modeling seeking to quantify the timing of the runoff peak and/or the shape of the overall runoff hydrograph resulting from a rainfall event. If the CN poorly models the runoff volume, the subsequent modeling of the peak discharge will be poor (Moglen et al., 2022).

But, the initial CN technique fails to include the slope of the terrain explicitly, which is a crucial factor for the behavior of runoff, particularly in hill-prone regions such as Uttarakhand. The steep slopes lower infiltration rates and raise the rate of overland flows, thus, enhancing the creation of runoff. Various research studies have tried to overcome this shortcoming. A slope-adjusted CN formula was offered by Huang et al. (2006) that will change the CN according to terrain steepness. Their results indicated that the inclusion of slope contributed immensely towards the accuracy of predicting runoff on watersheds that were sloping. The present study uses the CN approach as a point of reference in estimating the runoff in Uttarakhand area. To do this, first runoff is calculated as usual without slope and using standard CN values. Thereafter, the effects of the slope are considered within the model and CN values are adjusted according to slope steepness as provided by the literature adjustment methods (Huang et al., 2006). The implementation methodology, using ArcMap, targets the combination of LULC, HSG, rainfall and slope rasters to project the potential maximum retention (S), initial abstraction (I_a) and direct runoff (Q).

This research also examines the usefulness of Low Impact Development (LID) practices in reaction to the growing fears of urban runoff and extent of hard surfaces. LID is a collection of design measures employing either natural systems or human fabricated small-scale hydrological controls to cut the quantity of runoff and enhance the quality of water (Luan et al., 2017). The technologies that have been used to do this include permeable pavements, green roofs, bioswales and rain gardens and the objectives are to improve infiltration and reproduce pre-development hydrology. To achieve this study, LID is artificially applied by lowering CN values in urban and agriculture areas that indicate the increased infiltration abilities.

This chapter focuses on implementation of LID techniques to reduce runoff. These techniques reduce peak runoff, enhance infiltration and promote sustainable land use planning, LIDs serve as an effective response to increasing imperviousness due to urban expansion. Various LID interventions are simulated and integrated within the runoff model to assess their potential impact on reducing runoff volumes. The results demonstrate how incorporating slope and adopting LID strategies can support better runoff management, particularly in regions like Uttarakhand that have rugged topography and rapid development (Arjenaki et al., 2021).

2 Study Area

The study is conducted on the Indian state of Uttarakhand, a region characterized by diverse topography which ranges from low-lying foothills to steep mountainous terrains of the Himalayas. It is located in the northern part of India and lies between 28°44′ and 31°28′ N latitudes and 77°35′ and 81°01′ E longitudes (Chauhan et al., 2018). The region covers an area of around 53,483 km^2, comprising of 13 districts, many of which are prone to frequent hydrological events like landslides, flash flood and high surface runoff.

Uttarakhand's complex terrain, high rainfall variability and rapid urbanization make it a suitable and critical region for hydrological modelling. The presence of steep slopes, dense forests, agricultural zones, and expanding urban settlements creates dynamic land use and land cover patterns, all of which significantly influence surface runoff generation. Due to the fragile mountainous ecosystem, sustainable runoff management becomes essential to mitigate flood risks, control soil erosion, and maintain water quality (Ghosh et al., 2019).

Another important climatic feature of the area is that there is a humid subtropical to temperate climate with the main portion of precipitation being during months of the monsoons. The amount of rain that is received is highly variable with elevation and location, usually between 1,200 mm and more than 2,000 mm of annual rainfall; thus, runoff estimation and management are major concerns for planners and environmentalists.

This research centers on using the Curve Number (CN) technique to calculate runoff in different slope scenarios and the incorporation of LID practices in specified areas of Uttarakhand to reduce runoff in the chosen areas. The variety in physiography within the state can provide useful comparison when it comes to the evaluation of the effectiveness of LID techniques under mountainous conditions (Fig. 1).

3 Methodology

The approach used in this research is based on LID methodology to demonstrate the usefulness of these Low Impact Development Techniques in the management of the surface runoff in the area of interest. The Curve Number method created by the United States Department of Agriculture Natural Resources Conservation Service (USDA-NRCS) was used to quantify the runoff (U.S. Department of Agriculture & Natural Resources Conservation Service, 2004). The technique was used in two steps

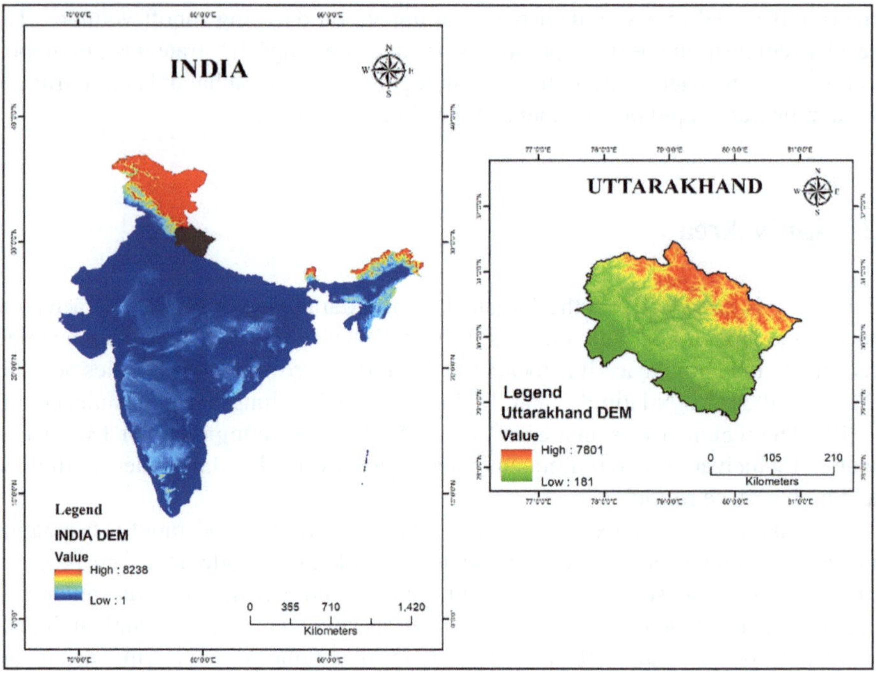

Fig. 1 Study area

namely first without slope inclusion and then with slope to determine the effects it has on runoff estimates.

3.1 Datasets

3.1.1 Satellite Imagery

The United States Geological Survey (USGS) Earth Explorer was used to acquire the satellite imagery for 2004, 2014, and 2024 (https://earthexplorer.usgs.gov). Images were downloaded at a spatial resolution of 30 m, and Landsat imagery was chosen based on the least amount of cloud cover. To produce precise land use/land cover (LULC) layers for every year, supervised classification was carried out in ERDAS Imagine rather than using pre-classified LULC maps. This made it possible to depict changes in land cover that affect surface runoff in great detail.

3.1.2 Digital Elevation Model (DEM)

Shuttle Radar Topography Mission (SRTM) DEM was used with a spatial resolution of 30 m. It was acquired from USGS Earth Explorer (https://earthexplorer.usgs.gov). It was then processed in ArcMap software to derive slope maps, which were later incorporated into CN estimation.

3.1.3 Rainfall Data

Data for 2004, 2014 and 2024 was downloaded from Indian Meteorological Department (IMD) in Network Common Data Form (NetCDF) (https://imdweb.lwcc.in/) (Nandi et al., 2023). These files were opened in ArcGIS 10.8 software and the data was extracted using the Multidimension Tools and converted it into useful raster format. Monthly rainfall values were averaged to derive annual precipitation, which was then used as input in the SCS-CN runoff calculation model.

3.1.4 Hydrological Soil Group Data

The Harmonized World Soil Database (HWSD) provided the hydrological soil group (HSG) data which is a classification of soil in terms of infiltration capacity; this is accessible through both FAO and IIASA. Maps were generated by adding the HSG data and LULC maps for each year. Each map was then reclassified according to the Curve Number of the particular class.

3.1.5 Curve Numbers

The Curve Numbers were referenced from USDA Natural Resources Conservation Service (NRCS) Curve Number Table, which was published in a book (U.S. Department of Agriculture & Natural Resources Conservation Service, 2004). Publications from the Soil Conservation Service (1971, 1986) provide further background and details on use of the CN model.

Arc GIS 10.8 software was used for the analysis. All the datasets were projected to WGS 84 and UTM Zone 43 for spatial consistency.

3.2 Direct Runoff Generation

The CN approach is based on two fundamental principles and the water balance equation. The first hypothesis contrasts the ratio of actual direct surface runoff (Q) to total precipitation (P) (or maximum potential surface runoff) (S) with the ratio of prospective maximum retention (P) to actual maximum infiltration (F) because the second hypothesis connects potential maximum retention (PMR) and initial abstraction (I_a). CN method estimates direct runoff using the formula (Karunanidhi et al., 2020):

$$Q = \frac{(P - I_a)^2}{(P - I_a) + S} \quad \text{for } P > I_a, \tag{1}$$

$$Q = 0 \quad \text{for } P \leq I_a, \tag{2}$$

where Q is the direct runoff (mm), P is the rainfall (mm), S is the potential maximum retention (mm) and I_a is the initial abstraction.

3.2.1 Land Use Land Cover (LULC) Map

LULC analysis is a general term used to depict Earth surface cover, whether it is natural or manmade. According to the United Nations' Food and Agricultural Organization (FAO, 1999), land cover is the observed physical and biological cover of the Earth's land, as vegetation or man-made features. Land use is the total of arrangements, activities, and inputs that people undertake in a certain land-cover type (Environmental Management, 2017). Satellite imageries of 2004, 2014, and 2024 were used to generate the LULC Maps. This was done using ERDAS Imagine 2021 software. Radiometric and geometric errors were corrected from the imageries. Around 100–150 training samples were taken to perform Supervised Classification. To ensure the accuracy of the classification, accuracy assessment was performed using Google Earth. An accuracy of about 80–82% was calculated for the maps through this.

3.2.2 Composite of LULC + HSG Maps

For hydrological study, the HSGs are formulated to understand the interaction of rainfall with land use and soil type. HSGs are mainly categorized into four groups depending upon their infiltration and runoff capacity. It varies from class A to class D where A represents high infiltration and low runoff potential. As it progresses from A to D, runoff potential increases while infiltration capacity of soil decreases. Class B represents the moderate fine to coarse textures and are considered as moderate infiltration rates. Class C is considered for fine textures with slow infiltration rate whereas D stands for clay soils having slow infiltration rates (United States Department of Agriculture, 1986). This analysis is performed in GIS environment where the soil map has been further categorized to HSGs. Furthermore, the HSG is combined with the land use classification map to generate a runoff curve map using established Curve Number tables (US Army Corps & of Engineers, 2023; Table 1).

Table 1 Reclassified HSG values

S. No.	HSG class value	Reclassified value
1	A	0
2	B	1
3	C	2
4	D	3

Table 2 Reclassified LULC values

S. No.	LULC class value	Reclassified value
1	Snow	100
2	Water bodies	200
3	Forest	300
4	Agriculture	400
5	Built up	500
6	Barren land	600

This created a reclassified HSG raster. Similarly, LULC maps were also reclassified as the HSG map. The values for LULC were reclassified as in Table 2.

3.2.3 Curve Number (CN) Map

The Curve Number (CN) values range from approximately 30 for permeable soils with high infiltration rates to 100 for water bodies and soils with low infiltration rates (United States Department of Agriculture & Natural Resources Conservation Service, 2004). A higher CN indicates a higher runoff potential and vice versa. The LULC + HSG maps were further reclassified according to the appropriate Curve Number value corresponding to the particular class using the Curve Number Table (U.S. Department of Agriculture & Natural Resources Conservation Service, 2004). Values from 100 to 603 were reclassified using the Spatial Analyst tool. With this a new map, Curve Number Map was generated with values ranging from 36 to 98. CN map was generated for each year using the same method.

3.2.4 Calculation of Potential Maximum Retention (S)

The potential maximum retention (S) represents the maximum amount of water (in mm) that can be retained by the soil and vegetation after runoff begins. It relates directly to CN, which is a dimensionless value ranging from 0 to 100, representing the runoff potential of a land surface (Barker & Chandrakantha, 2022). The potential maximum retention S (in millimeters) is calculated using the formula:

$$S = \frac{25400}{\text{CN}} - 254. \tag{3}$$

The constant 25,400 is derived by converting the original imperial-unit-based equation: $S = \frac{1000}{\text{CN}} - 10$, which yields S in inches, into millimeters. Since 1 inch equals 25.4 mm, multiplying both constants by 25.4 gives:

$$S = \left(\frac{1000}{\text{CN}} - 10\right) \times 25.4 = \frac{25400}{\text{CN}} - 254.$$

This conversion makes the formula suitable for hydrologic modelling in countries using the metric system.

3.2.5 Calculation of Initial Abstraction (I_a)

The initial abstraction (I_a) can be considered the boundary between the storm size that produces runoff and the storm size that produces no runoff (United States Department of Agriculture & Natural Resources Conservation Service, 2004). These include water retained in surface depressions, water intercepted by vegetation, evaporation, and infiltration.

$I_a = \lambda S$, here λ is the initial abstraction coefficient. $\lambda = 0.2$ was assumed in the original CN method. The 0.2 factor remains the standard in Curve Number methodology due to its practicality and historical validation (Barker & Chandrakantha, 2022).

$$I_a = 0.2 \times S. \tag{4}$$

This empirical relationship simplifies the calculation of direct runoff.

3.2.6 Direct Runoff Estimation

Direct runoff estimation refers to the process of calculating the volume or rate of water flowing over the land surface and into streams, rivers, or other bodies of water, specifically after a rainfall event, and after accounting for initial losses like infiltration and depression storage (Basak et al., 2017). Equation (1) is used here and all the values obtained from Eqs. (3) and (4) are substituted in Eq. (1). Hence, direct runoff estimation is collected.

3.3 Direct Runoff Estimation Incorporating Slope

For calculating direct runoff with slope, same steps are followed as in direct runoff estimation, the only change is in the CN map. After generating the CN map, a slope-adjusted CN map is created. To calculate this, Slope percentage (S) and CN values are needed. The formula for this is as follows (Huang et al., 2006):

$$\mathrm{CN_{adj}} = \mathrm{CN} \times \left(1 + \frac{S}{100}\right). \quad (5)$$

3.4 Impact of LID Techniques on Direct Runoff

A variety of LID practices have been developed to address the problem of runoff and improve ecosystem services. Some common practices are bioretention cells that involve use of vegetation and soil to gather runoff, vegetated swales, replacing concrete gutters, green roofs, which reduce runoff in addition to serving as insulation and beautiful facades, and permeable pavement, which facilitate infiltration of the water in urban areas (Sui & van de Ven, 2023). To show the impact of LID techniques, the initial CN values were reduced for the land use type where these techniques were being implemented. CN values of agriculture, built-up and barren land were manipulated and reduced as per LID techniques implementation.

4 Results and Discussion

4.1 Land Use Land Cover Classification

Land Use Land Cover is used for sustainable management of the environment. The LULC classification for the years 2004, 2014, and 2024 was carried out using ERDAS Imagine software with a supervised classification approach. The classification distinguished six major classes: snow, water bodies, forest, agriculture, built-up, and barren land. Supervised Classification with maximum likelihood algorithm was applied, followed by accuracy assessment to ensure classification reliability. The spatial extent of each LULC category was computed, revealing notable changes across the three decades (Bhuvaneswari et al., 2020; Table 3 and Fig. 2).

Table 3 Area distribution of land use/ land cover (LULC) classes in Uttarakhand

S. No.	Class name	Area (in km^2)
1	Snow	8,171.27
2	Forest	3553.44
3	Water bodies	1195.03
4	Agriculture	2677.24
5	Built up	2062.01
6	Barren land	8343.84

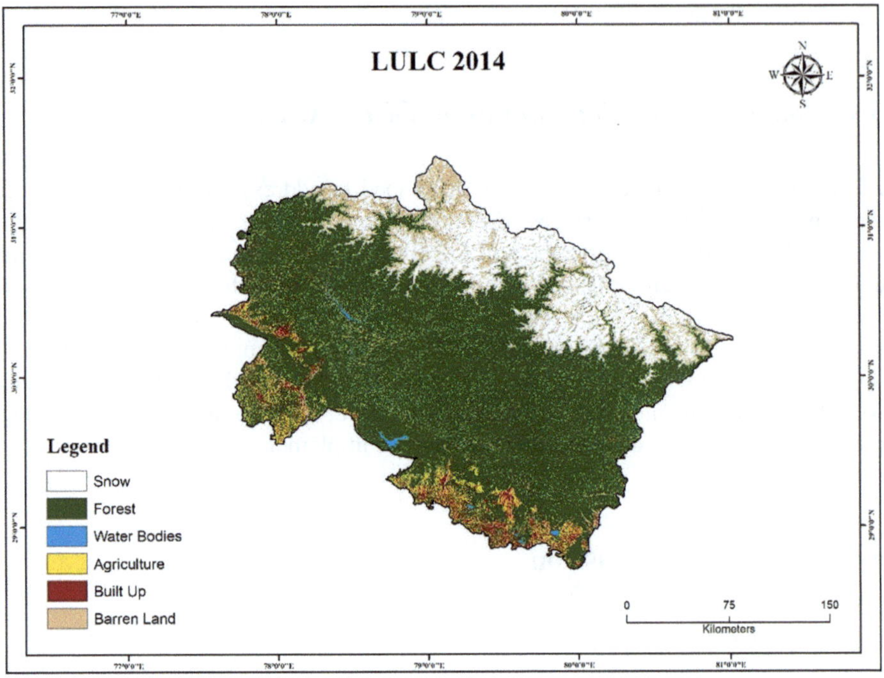

Fig. 2 LULC classification of Uttarakhand, 2014

4.2 Hydrological Soil Group

HSG is one of the main components used to calculate the Curve Number (CN). Hydrological soil groups were identified for the entire state of Uttarakhand based on the USDA-Soil Conservation Service (SCS) guidelines to determine the rainfall-surface runoff. The HSG classification of soil mainly depends on the infiltration rate of each soil texture category and was grouped into four categories A, B, C, and D. Forested and undisturbed areas were largely dominated by A and B groups, whereas built-up and barren areas corresponded to C and D soils due to reduced infiltration potential (Table 4 and Fig. 3).

Table 4 Area distribution of hydrological soil groups (HSGs) in Uttarakhand

S. No.	Soil group name	Area (in km^2)
1	A	5.782
2	B	4.968
3	C	6.967
4	D	12.347

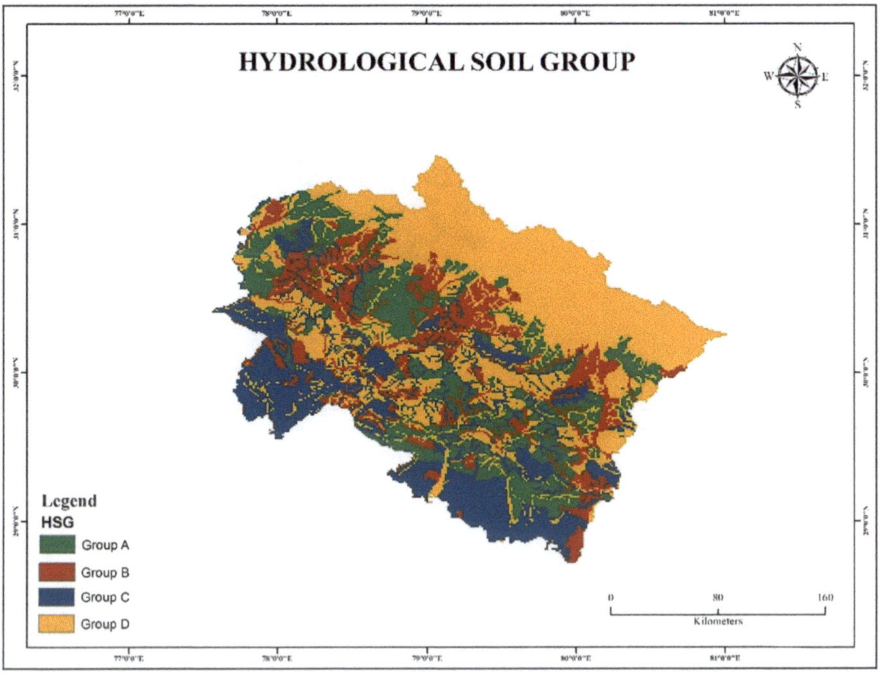

Fig. 3 Hydrological soil group (HSG)

4.3 Composite LULC + HSG Map

This composite map has been generated to provide a spatially integrated view of land use characteristics and soil hydrologic properties. This map will be used further as an input for the CN map. The LULC classes include snow, water bodies, forest, agriculture, built-up and barren land while HSG categories (A, B, C, and D) represent soils with high to very low infiltration capacities. The composite representation highlights the interaction of these two factors (Fig. 4).

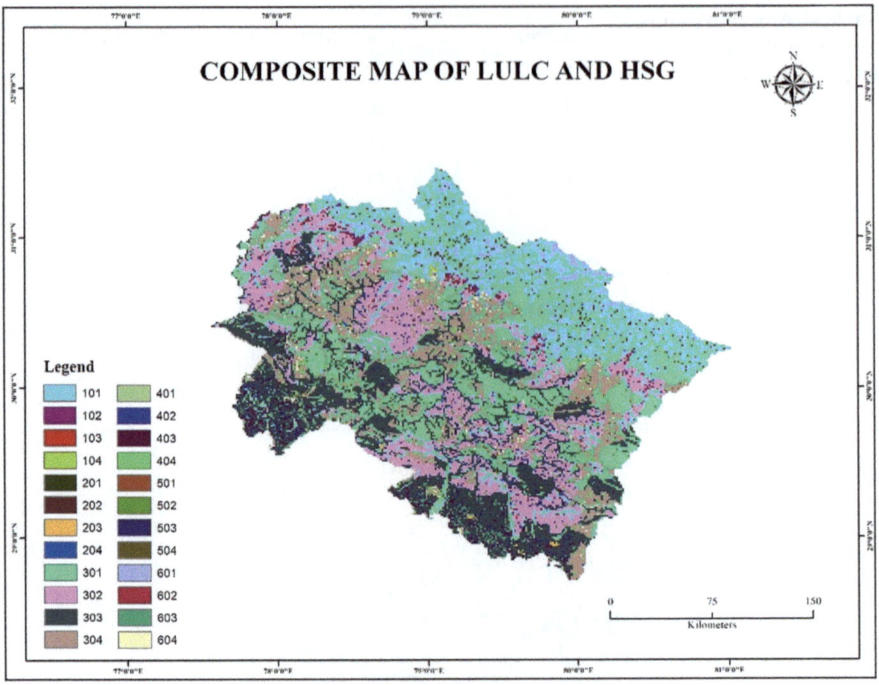

Fig. 4 Composite map of LULC + HSG of Uttarakhand, 2014

4.4 Curve Number Map

This map is prepared by integrating LULC and HSG data. CN values represent the runoff potential for different regions and may vary between 36 and 98 in the study area. Lower CN values correspond to regions with dense forest cover, permeable soils, and high infiltration capacity, resulting in lesser runoff. In contrast, higher CN values (above 90) are observed in built-up and barren areas where impervious surfaces or compacted soils significantly reduce infiltration, leading to higher runoff generation (Fig. 5).

4.5 Potential Maximum Retention (S)

S values expressed in millimeters, were calculated from CN values using Eq. (3). This equation relates CN to the maximum potential retention after runoff begins. Regions with high CN values yielded low *S* values, indicating limited capacity for infiltration and retention, whereas forecasted and permeable soils exhibited higher *S* values (Fig. 6).

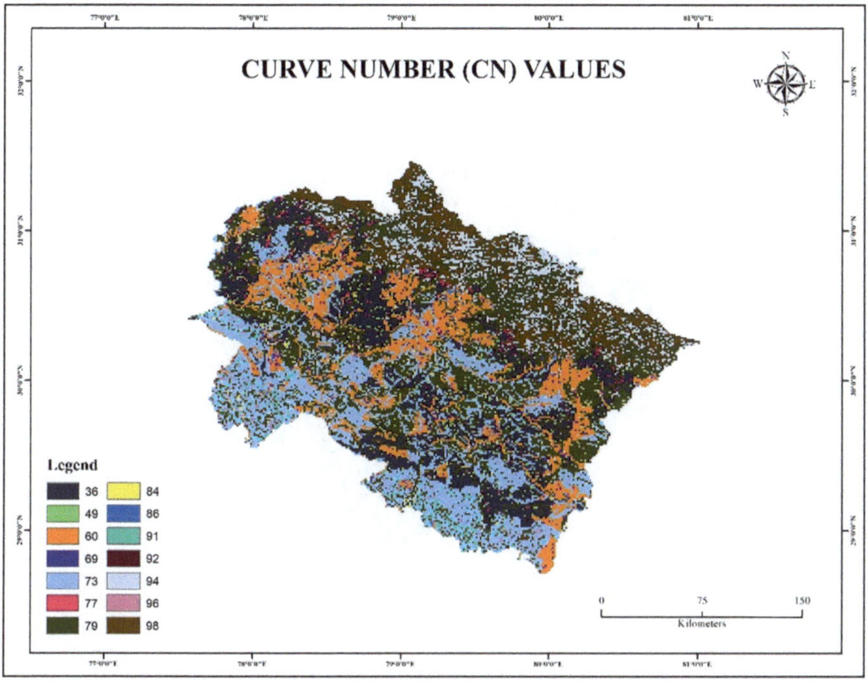

Fig. 5 Curve number map of Uttarakhand, 2014

4.6 Initial Abstraction (I_a)

Initial abstraction, including interception, infiltration, before runoff, and surface storage, was computed using Eq. (4). This stage identified areas where a significant portion of rainfall was absorbed before generating runoff. Permeable and vegetated areas showed higher I_a values (Fig. 7).

4.7 Rainfall

Rainfall data was obtained from Indian Meteorological Department (IMD) in Network Common Data Form (NetCDF) and spatially interpolated using Inverse Distanec Weighting (IDW) in ArcMap to generate a rainfall distribution map. The spatial variation revealed higher precipitation in the windward Himalayan slopes and reduced rainfall in leeward and valley regions. Average annual rainfall values ranged from 478.53 mm in low-rainfall zones to 2858.63 mm in high-rainfall zones. These variations significantly influenced the runoff patterns, as regions with high rainfall coincided with steep slopes and high CN values, resulting in amplified direct runoff (Fig. 8).

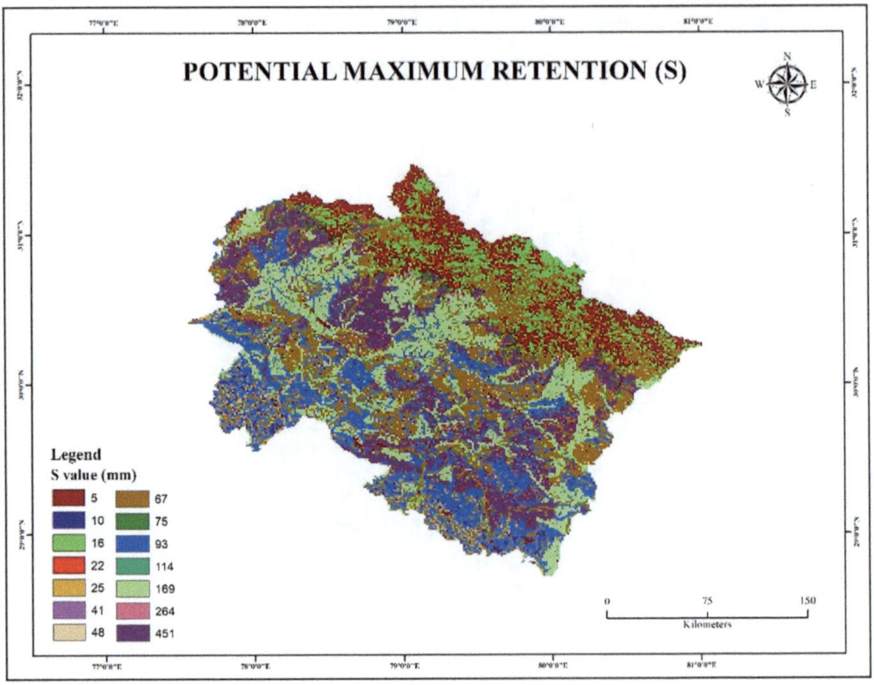

Fig. 6 Potential maximum retention (*S*) of soil of Uttarakhand, 2014

4.8 Direct Runoff

Direct runoff (Q) was calculated for each pixel using Eq. (1). Runoff was highest in built-up and barren areas due to high CN values and low infiltration. Agricultural land displayed moderate runoff, which varied depending on crop cover and soil permeability. Forested areas showed comparatively low runoff, reflecting higher infiltration and water retention. Water bodies, as expected, contributed minimally to direct runoff estimation as they primarily act as sinks (Fig. 9).

4.9 Extraction of Slope Percentage from DEM

The slope was derived from Digital Elevation Model (DEM) using the slope function in ArcMap. The slope values were expressed in percentage and classified into categories ranging from gentle slopes in the valleys to very steep slopes in the higher Himalayan region. The results show that a significant portion of the state is characterized by steep and very steep slopes, reflecting the rugged mountainous topography. These steep slopes directly influence runoff behavior, soil erosion potential, and land use practices, as they restrict agricultural expansion and settlement. Conversely,

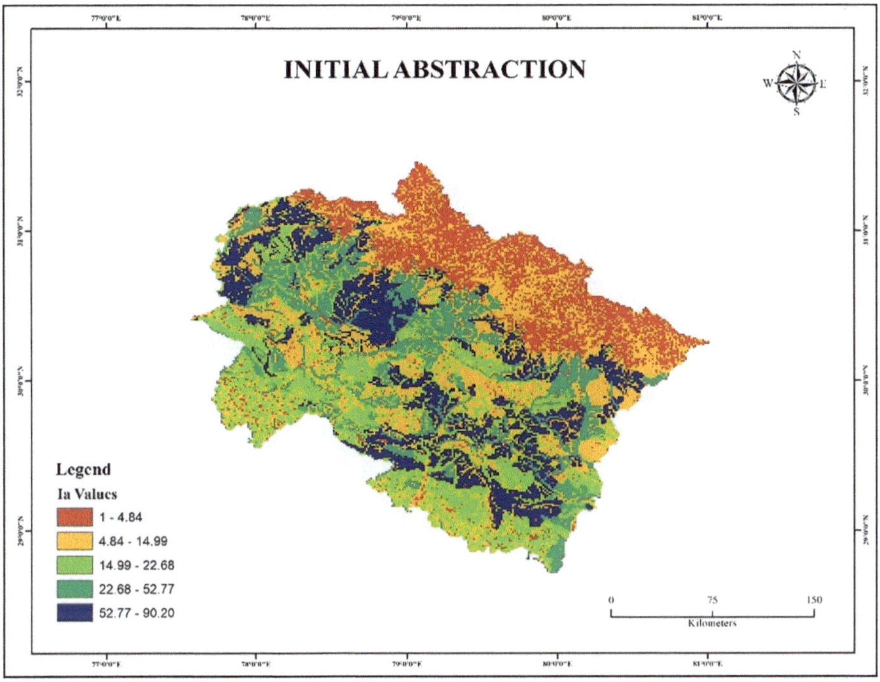

Fig. 7 Initial abstraction (I_a) of Uttarakhand, 2014

gentler slopes are concentrated in the plains and valleys, where agriculture and built-up areas are more dominant (Fig. 10).

4.10 Slope Adjusted Runoff

To account for the influence of topography on runoff generation, the slope map was integrated with the initial direct runoff estimates. Using slope-adjusted factors, areas with higher slope gradients exhibited greater runoff potential due to faster surface flow and limited infiltration. The slope-adjusted runoff map clearly highlights zones of high hydrological sensitivity, particularly in steep mountainous terrains, where the combination of high rainfall and slope enhances runoff generation. Comparatively, the flatter regions showed reduced runoff potential, with more opportunities for infiltration and storage. This adjustment provided a more realistic estimation of surface runoff across the diverse physiographic zones of Uttarakhand, improving the accuracy of hydrological modelling and site suitability analysis (Fig. 11).

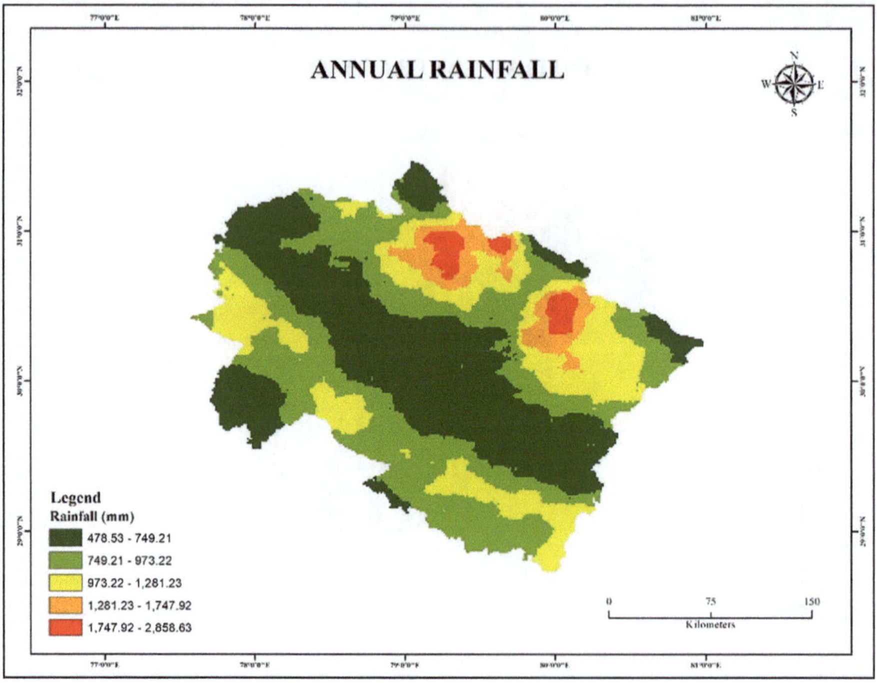

Fig. 8 Annual rainfall (*P*) of Uttarakhand, 2014

4.11 Runoff After the Implementation of LID Techniques

LID techniques, including bioretention cells, vegetated swales, permeable pavements, and green roofs, were simulated to evaluate their effectiveness in reducing runoff. These techniques were applied to high-runoff zones identified in the slope-adjusted Qmap. Post LID simulation results demonstrated a significant decrease in runoff volume, particularly in urban and semi-urban areas. These results show that although visual differences among the maps are slight, LID techniques produce a cumulative effect in the attenuation of runoff peaks and the redistribution of surface water to lower runoff classes. While the magnitude of attenuation is small at a catchment scale, it is important for flood mitigation, groundwater replenishment, and sustainable watershed management in a hydrologically sensitive area such as Uttarakhand (Fig. 12).

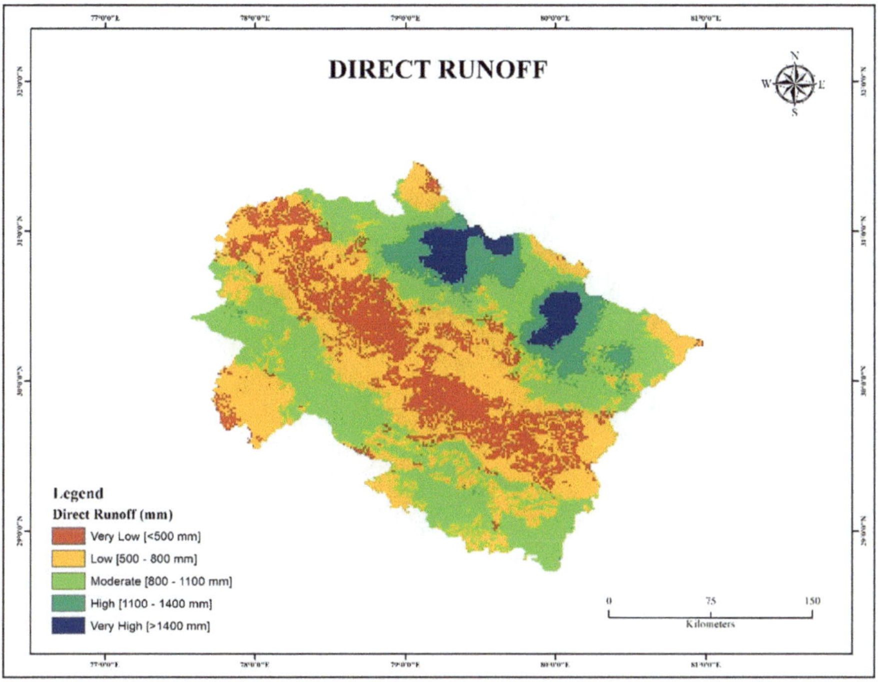

Fig. 9 Direct runoff estimation (Q) of Uttarakhand, 2014

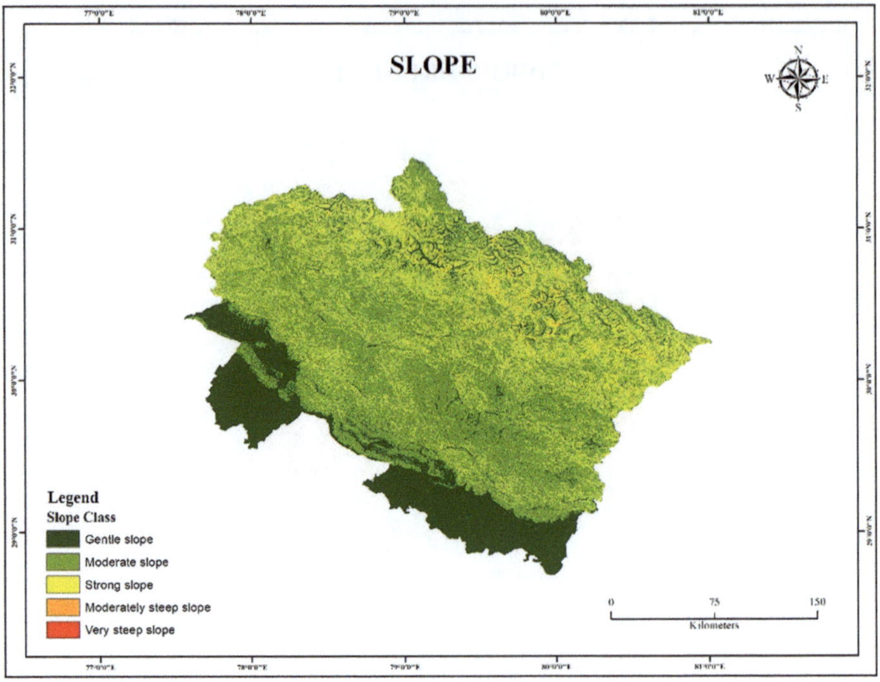

Fig. 10 Slope of Uttarakhand

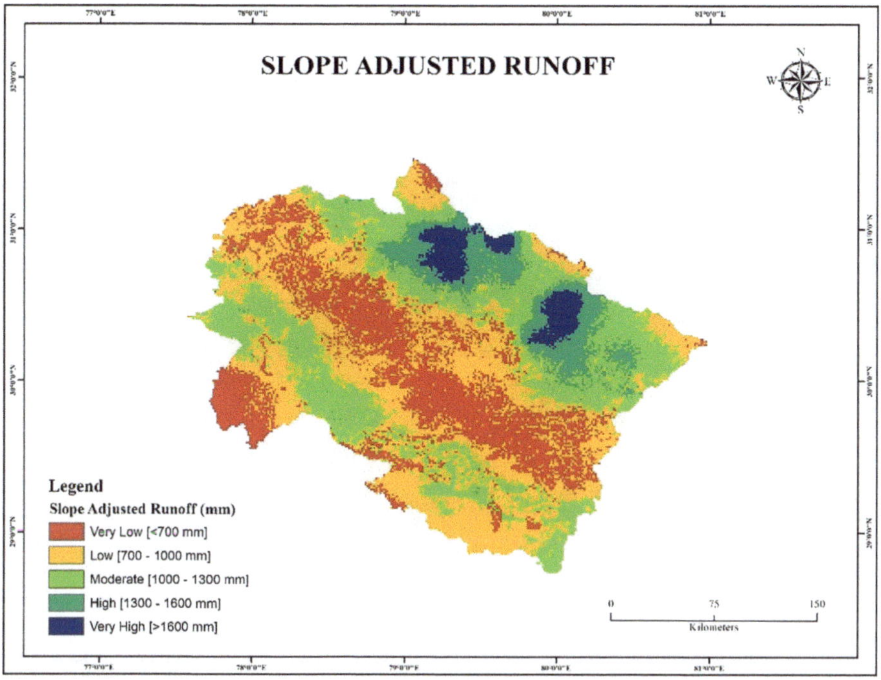

Fig. 11 Direct runoff estimation (*Q*) incorporating slope of Uttarakhand, 2014

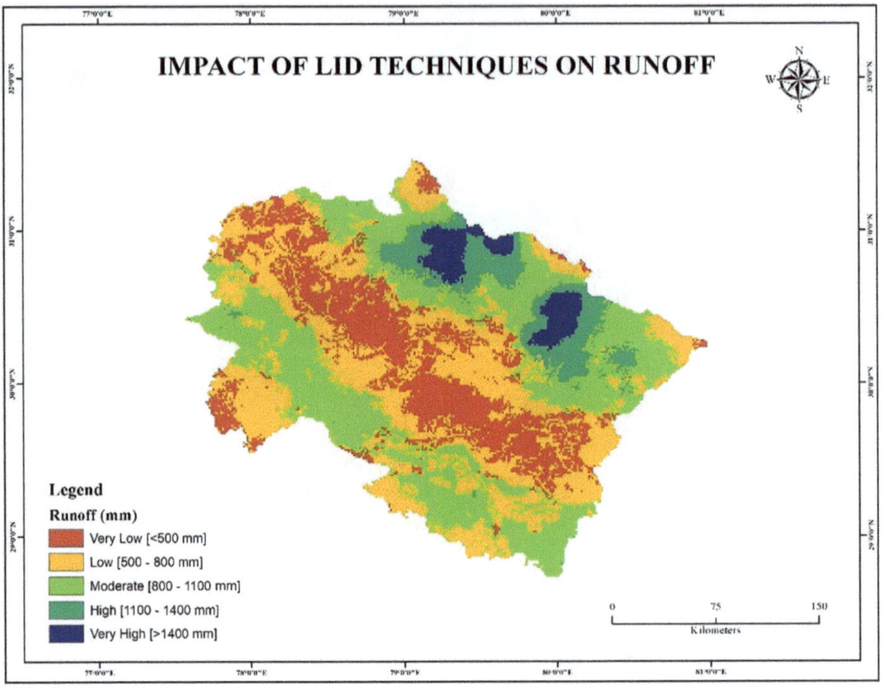

Fig. 12 Impact of LID techniques on runoff of Uttarakhand, 2014

5 Conclusion

The pairing of LULC dynamics and hydrological processes is of great importance for assessing runoff behavior in mountainous areas. In this study, the classification of LULC into snow, water bodies, forest, agriculture, built-up, and barren land in combination with rainfall variability provided useful information to analyze the surface runoff response (Tankpa et al., 2021). The Curve Number (CN) method was used to estimate the direct runoff (Q) with parameters such as the initial abstraction (I_a) and the potential maximum retention (S). Results indicated that built-up and agricultural lands contributed much to increased runoff, while forested and barren areas had relatively low runoff responses. These results indicate that spatial heterogeneity in CN values plays an important role in determining the distribution of rainfall excess and its conversion into direct runoff (Bhuvaneswari et al., 2020). The analysis also highlights the role of infiltration promoting land covers in reducing peak flows and flood hazards. By correlating LULC change with hydrological models, this research contributes a scientific basis for sustainable water resource planning and better management of runoff in the Himalayan environment (Garen & Moore, 2005).

References

Arjenaki, M. O., Sanayei, H. R. Z., & Heidarzadeh, H. (2021). Modeling and investigating the effect of the LID methods on collection network of urban runoff using the SWMM model (case study: Shahrekord City). *Modeling Earth Systems and Environment, 7*, 1–16. https://doi.org/10.1007/s40808-020-00870-2

Barker, R. D., & Chandrakantha, G. (2022). Estimation of runoff potential using Curve Number Method. *Indian Journal of Ecology, 49*(5), 1843–1847. https://doi.org/10.55362/IJE/2022/3745

Basak, D., Islam, M., Islam, M. T., & Shariot-Ullah, M. (2017). Estimation of direct runoff and runoff coefficient for different storm events in a small catchment.

Bhuvaneswari, K., Geethalakshmi, V., Lakshmanan, A., Srinivasan, R., & Sekhar, N. U. (2020). Rainfall–surface runoff estimation for the Lower Bhavani basin in south India using SCS-CN model and geospatial techniques. *Environmental Earth Sciences, 79*(335). https://doi.org/10.1007/s12665-020-09077-4

Chauhan, J. S., Gautam, A. S., & Negi, R. S. (2018). Natural and anthropogenic impacts on forest structure: A case study of Uttarakhand state. *The Open Environmental Research Journal, 11*, 38–46. https://doi.org/10.2174/1874213001811010038

Environmental Management. (2017). *Land cover analysis*. Elsevier. Accessed July 20, 2025. https://sciencedirect.com/topics/earth-and-planetary-sciences/land-cover-analysis

Garen, D. C., & Moore, D. S. (2005). Curve number hydrology in water quality modeling: Uses, abuses, and future directions. *Journal of the American Water Resource Association, 41*(2), 377–388. https://doi.org/10.1111/j.1752-1688.2005.tb03742.x

Garg, V., Nikam, B. R., Thakur, P. K., & Aggarwal, S. P. (2013). Assessment of the effect of slope on runoff potential of a watershed using NRCS-CN method. *International Journal of Hydrology Science and Technology, 3*(2), 141–159. https://doi.org/10.1504/IJHST.2013.057626

Ghosh, T. K., Jakobsen, F., Joshi, M., & Pareta, K. (2019). Extreme rainfall and vulnerability assessment: Case study of Uttarakhand rivers. *Natural Hazards, 99*, 665–687. https://doi.org/10.1007/s11069-019-03758-6

Huang, M., Gallichand, J., Wang, Z., & Goulet, M. (2006). A modification to the Soil Conservation Service curve number method for steep slopes in the Loess Plateau of China. *Hydrological Processes, 20*(3), 579–589. https://doi.org/10.1002/hyp.5925

Kansal, M. L., & Singh, S. (2022). Flood management issues in hilly regions of Uttarakhand (India) under changing climatic conditions. *Water, 14*(12), 1879. https://doi.org/10.3390/w14121879

Karunanidhi, D., Anand, B., Subramani, T., & Srinivasamoorthy. K. (2020). Lower Bhavani basin in south India using SCS–CN model and geospatial techniques. *Environmental Earth Sciences, 79*(335). https://doi.org/10.1007/s12665-020-09079-z

Luan, Q., Fu, X., Song, C., Wang, H., Liu, J., & Wang, Y. (2017). Runoff effect evaluation of LID through SWMM in typical mountainous, low-lying urban areas: A case study in China. *Water, 9*(6), 439. https://doi.org/10.3390/w9060439

Moglen, G. E., Sadeq, H., Hughes II, L. H., Meadows, M. E., Miller, J. J., Ramirez-Avila, J. J., & Tollner, E. W. (2022). NRCS Curve Number Method: Comparison of methods for estimating the curve number from rainfall-runoff data. *Journal of Hydrologic Engineering, 27*(10). https://doi.org/10.1061/(ASCE)HE.1943-5584.0002210

Nandi, S., Patel, P., & Swain, S. (2023). IMDLIB: An open-source library for retrieval, processing and spatiotemporal exploratory assessments of gridded meteorological observation datasets over India. *Environmental Modelling & Software, 167*, Article 105869. https://doi.org/10.1016/j.envsoft.2023.105869

Sui, X., van de Ven, F. H. M. (2023). The influence of Low Impact Development (LID) on basin runoff in a half-urbanized catchment: A case study in San Antonio, Texas. *Journal of Hydrology, 616*, 128793. https://doi.org/10.1016/j.jhydrol.2022.128793

Tankpa, V., Wang, L., Awotwi, A., Singh, L., Thapa, S., Atanga, R. A., & Guo, X. (2021). Modeling the effects of historical and future land use/land cover change dynamics on the hydrological response of Ashi watershed, northeastern China. *Environment, Development and Sustainability, 23*(6), 7883–7912. https://doi.org/10.1007/s10668-020-00939-y

United States Department of Agriculture. (1986). *Urban Hydrology for Small Watersheds*. Technical Release 55 (TR-55). Natural Resources Conservation Service, Conservation Engineering Division.

U.S. Department of Agriculture, Natural Resources Conservation Service. (2004, July). Chapter 9: Hydrologic soil-cover complexes. *National engineering handbook: Part 630—Hydrology*. USDA NRCS.

United States Department of Agriculture, Natural Resources Conservation Service. (2004). Chapter 10: Estimation of direct runoff from storm rainfall. *Part 630 hydrology: National engineering handbook*. USDA NRCS. https://irrigationtoolbox.com/NEH/Part630_Hydrology/H_210_630_10.pdf

US Army Corps of Engineers. (2023). *HEC-RAS hydraulic reference manual. Version 6.7*. Hydrologic Engineering Center. Accessed July 20, 2025, from: https://www.hec.usace.army.mil/confluence/rasdocs/ras1dtechref/latest/overview-of-optional-capabilities/modeling-precipitation-and-infiltration/curve-number

Reviving Traditional Water Wisdom of India: A Sustainable Pathway for Socio-ecological Resilience

Alpna, Fatima Amin, and Anil Kumar Gupta

Abstract Revival of traditional water wisdom offers a transformative pathway to enhance socio-ecological resilience under growing water scarcity, ecological degradation and climate change. India has centuries-old indigenous community-driven systems rooted in sustainability, equity, and adaptation to diverse geo-climatic conditions. However, due to rapid urbanization, policy neglect and centralized infrastructure, these practices get eroded. This chapter explores the traditional water management and practices related to water conservation, irrigation and water productivity for fostering social cohesion, cultural values and ecological balance. Such models illustrate integration of traditional wisdom into urban and rural planning can strengthen water resource efficiency, diversify sources, and reduce dependency on energy-intensive supply systems. The chapter also includes a roadmap for policy recognition and integration, community participation, and knowledge-sharing platforms to bridge generational wisdom with current needs. Thus, by revaluation of traditional water wisdom, societies can unlock cost-effective, localized and sustainable solutions, reduce disaster risks, and support climate-resilient development, ensuring resilience and adaptive capacity for present and future challenges.

Keywords Traditional water wisdom · Socio-ecological resilience · Sustainability · Policy landscape

1 Introduction

India has a rich history of water conservation and management dating back to ancient civilizations like the Harappa and Vedic periods (Singh et al., 2020). The perspective of hydro-science, hydro-mechanics and water management in ancient Indian

Alpna · F. Amin (✉) · A. K. Gupta
Integrated Centre for Adaptation, Disaster Risk Resilience and Sustainability (ICARS), IIT Roorkee, Greater Noida Campus, Uttar Pradesh, India

Alpna
e-mail: alpna.icars@gmail.com

S. Pathak and M. Pandey (eds.), *Geospatial Technology and Advanced Instrumentation in Addressing Water Resources Planning and Management*, Water Informatics for Water Resource Management, https://doi.org/10.1007/978-981-92-1602-4_7

scriptures like Rig-Veda, Yajur-Veda, Atharv-Veda, Puranas i.e. Ramayana, Mahabharata and other texts such as Brihat-Samhita, Arthashastra and Meghamala provide invaluable verses that cover topics such as the water cycle, formation of clouds, flow of streams, level of underground water, drought management, method of watering crops, arrangement of ponds and protection of bank measures including plantation with pitching and sluicing, hydraulic machinery, different hydro-structures and eco-friendly approaches for managing wastewater (Dasgupta et al., 2024). However, many of these practices declined post-independence due to negligence and policy changes (Jain et al., 2024).

In recent times, due to population of over a billion people, the erratic monsoon, land degradation, pollution and overexploitation of natural resources in India are leading the country towards social, economic, political and environmental crisis. There is an increasing recognition of value to traditional water management wisdom as an adaptable and environmentally sustainable solution for conserving natural resources, promoting social and ecological equity with reducing environmental impacts in given challenges of water crisis, environmental degradation, extreme hydro-meteorological events and changing climate. Adopting the traditional techniques can contribute to achieving Sustainable Development Goals (SDGs), namely, SDG 6, which addresses the clean water and sanitation, SDG 13 related to climate action and SDG 15 focusing on life on land. Researchers emphasize the need for policy reforms, increased community participation, and integration of modern technologies to enhance the efficiency of traditional water management structures in preventing and mitigating natural hazards and improving environmental support services (Jain et al., 2024; Singh et al., 2021). The chapter discusses the various ancient regional water management system related to water conservation, irrigation and water productivity with roadmap of their revival in present challenges.

2 Regional Ancient Water Management Systems

From the arid deserts of Rajasthan to the fertile Ganges-Yamuna plains and the coastal belts, India reflected an array of indigenous water system designed to suit, store and distribute water with harmony in nature. Many of these ancient systems, including Johads, Baolis, Tanks, Ahar-Pyne, Phad system, etc. were not only engineering and architectural marvel but also socio-ecological, religiously and culturally rooted in local governance and community engagement. The stewardship and revitalization of water bodies underline their significance in regulating heat and conserving biodiversity, which is integral to ecological harmony and climate resilience (Saini et al., 2025). Some of the most significant of these traditional systems are explained below, illustrating how ancient wisdom ensured sustainable water use for centuries.

2.1 Traditional Structures and Practices Related to Water Conservation

In India, various traditional water conservation systems are in use since centuries, i.e. roof water harvesting which facilitates groundwater recharge, cost-effective system for rejuvenation of depleting water resources, i.e. Ramtek model, and Integrated water resources management (IWRM) approach for water security. Table 1 shows some of the locally available traditional water conservation practices in different locations of India.

The following sections provide a brief overview of using traditional systems by local people for water security, groundwater recharge, water bodies revival and depleting river discharge in different regions of India.

Northern Region

Baudi is a circular or square shape deep stoned pit to facilitate continuous percolation of water naturally from earth surface. When its size increased and water uses for multiple purposes, it is called ***Nawn***. A stone-based tank like structure is constructed for locating the water source, from which different channel segregated for drinking water, wash clothes or take bath (Sharma & Kanwar, 2009). These both systems have provision of roofs with three side walls to protect the water. A ***Chappri/Talaais/Talaabs*** is a shallow unlined pond, found on hillside where the terrain flattens out for providing water to livestock and irrigation purposes. ***Khatris*** are made in hard rocks on the hilly terrain. These are rectangular shape deep pits where rainwater is collected through seepage from rocks and soils. The structure of Khatris is built by digging a horizontal tunnel of 3–4 m length, followed by a vertical basin at the inner end. Its capacity varies between 30,000 and 50,000 L. These traditional water storage systems are found in Hamirpur, Kangra and Mandi districts of Himachal Pradesh (Sharma & Kanwar, 2009). ***Chaal*** and ***Khal*** (artificial ponds to capture rainwater on hilltops), ***Naula*** (small stepwell) and ***Dhara*** (perennial springs) are incredible water-filled reservoirs of Kumaon division of Uttarakhand, which serve as a primary drinking water source, used in irrigation and power watermills. Naula is a small groundwater harvesting structure which is built by locally available stone wall with a stone roof, and water enter the stepwell via a channel or pipe from the source. These are 6–8 ft. at the top and inverted trapezoids at the bottom with a square of 1–1.5 ft. (Bhandari & Kaur, 2023). ***Dighis*** was first introduced by Mughal emperor Shahjahan (1627–58 AD) in Delhi to meet the water demand for domestic use. It is a square or circular reservoir with steps to enter, and these had its own sluice gates. Kahar or a Mashki were hired by people to draw water from it.

Western Region

Nadi is the village pond in Jodhpur, used for storing water during rainy season from a nearby natural catchment. After the rains, water is available in most of the Nadis ranging from 2 months to a year. This system delivers water for human and livestock purposes. It is estimated that the recharge from a Nadi spanning 2.25 hectares with a

Table 1 Traditional water conservation systems in India

Region	System	Practised in
For depleting groundwater		
North	Baudi	Himachal Pradesh
	Nawn	
	Dighi	Delhi
West	Paar system	Rajasthan
	Baolis/Baoris/Bawaris	
	Kuis/Beris	
	Dakeiyam	
	Jhalara	
	Kunds/Kundis	Gujarat, Rajasthan
South	Garanda	Karnataka
For depleting river discharge		
West	Ramtek model	Maharashtra
South	Jarukuttu	Karnataka
	Bavi/Kunta/Revu/Cheruvu	Andhra Pradesh and Telangana
For disappearing water bodies		
North	Chappris	Himachal Pradesh
Central	Chandela/Bundela tanks	Bundelkhand
	Talab/Bandhi/Sagar/Samand/Pokhariyan	
	Haveli	Madhya Pradesh
	Pemghara	Odisha
West	Johads	Rajasthan
	Nadi	
	Toba	
South	Madakas	Karnataka
For water security at village		
North	Chaal/Khal/Naula/Dhara	Uttarakhand
	Khatri	Himachal Pradesh
West	Tankas	Rajasthan
	Virdas	Gujarat
South	Surangam	Karnataka, Kerala
	Jack wells	A&N, Lakshadweep

storage capacity of 15,000 m^3 in an alluvial region can contribute about 10,000 m^3 of groundwater recharge in one rainy season (Bhattacharya, 2015). But in recent times, some of the factors e.g. water pollution, heavy sedimentation, significant evaporative and seepage losses are severely affecting Nadis (Saxena, 2017). ***Baoris/Stepwells*** are protected wells/ponds in which the water can be accessed by descending a series of steps. These are covered structure and may be multi-storied also. The step wells originated to secure the water needs during the period of drought in Gujarat and Rajasthan. ***Jhalara*** are rectangular shape structure and have steps on three or four sides. These are essentially meant for public use and religious practices in Rajasthan and Gujarat (Bhattacharya, 2015). ***Toba*** is similar to a Nadi, a ground depression with a natural catchment area for providing water to human and livestock consumption. The grass around the Toba provides pastures for livestock. These were deepened in the middle for concentrated storage and to minimized the evaporation. ***Kunds/ Kundis*** are the covered underground circular tank developed for handling drinking water crisis in western arid regions of Rajasthan. The Kunds have a saucer-shaped catchment area with a gentle slope towards the centre where a tank is situated. Usually constructed with native materials like lime plaster for tank, pond silt, murrum, charcoal ash and gravel for catchment areas. ***Kuis/Beris*** are 10–15 m deep pits dug next to tanks in order to collect their seepage. The mouth of the pit is usually very narrow to prevent evaporation and gets wider under the ground. ***Dakeiyam*** are shallow pits dug up in harvested fields to store rainwater for kharif cultivation (Saxena, 2017). ***Johads*** are small earthen check dams that trap and preserve rainwater, increasing infiltration, recharging groundwater, enhancing crop yields, lifting water table, intercept monsoon runoff and improving water availability. The water from Johads is used for drinking, livestock, irrigation and other various purposes. These are mud barriers in semi-circular shape built across hilly slope to trap the monsoon runoff, and bounded on three sides by the embankments while the fourth side is open for rainwater to enter. Johads are mainly found in the Thar desert of Rajasthan (Yadav, 2023). ***Paar*** is a rainwater harvesting system which has 5–12 m deep Kuis or Beris that are unearthed for the collection of percolated water (Rejani Pani) from the agar (catchment). Water collected using this method is also known as Patali Pani. ***Tankas*** are small underground tank mostly found in houses of Bikaner. These are generally circular holes lined with fine polished lime to collect rainwater for drinking purpose. However, ***Tanks*** are made of huge walls on four sides along with an impervious floor having enormous water holding capacity. ***Virdas*** are found in the Banni grasslands of the Great Rann of Kutch in Gujarat. These are shallow holes made in the sands of dry riverbeds and lakes for collecting water for drinking purpose. ***Ramtek model*** of Maharashtra includes sequence of surface and groundwater bodies in which water runoff is collected in Baoris, wells and Kundis.

Central Region

Chandela tanks were built by halting the water flow in rivulets flowing between hills by erecting massive earthen embankments. These tanks are made up of lime and mortar and provided drinking water for villagers and cattle. ***Bundela*** tanks are larger and had strongly constructed steps leading down to water in the tank. ***Haveli system***

of Madhya Pradesh is a remarkable cultivation structure based on water harvesting and runoff farming. Rainwater is stored in the fields with embankments until sowing season. As soon as field is sown, water is taken out and no further watering is needed (Gupta et al., 2014). ***Pemghara*** of Odisha are small earthen check dams built on high banks on three sides along with storage pits on fourth side. In some areas these are linked with deep channels that lead to a single outlet into the river or stream. They are called ***Madakas*** in Karnataka (Sundaram & Bhaskaran, 2021). A ***Pokhariyan*** is typically a small, natural pond, used for drinking and irrigation, while a ***Talab*** is a medium-sized lake, and a ***Bandhi*** is another term for a mid-size Lakes. Larger lakes or reservoirs are known as ***Sagar*** or ***Samand***.

Southern Region

Jackwells are bunds made from logs of hard bullet wood by Shompen tribe of Great Nicobar Island. Split bamboos placed along a gentle slope (placed below trees to collect rainwater through leaves) with bottom end draining water into a shallow pit. These act as channels for rainwater which direct it into pits called Jackwells (Bhattacharya, 2015). ***Surangam*** is a water harvesting system of northern Malabar region of Kerala. It is well dug in hard laterite rock, and water seeps out of rock and flows out of tunnel. This water is collected in an open pit constructed outside the Surangam. It is also known as Thurangam, Thorapu, Mala, etc. in different parts of Kasaragod district. ***Jarukattu*** is an embankment seems like the check dam. ***Cheruvu*** are found in Chitoor and Cuddapah districts in Andhra Pradesh. They are reservoirs to store runoff and are fitted with thoomu (sluices), alugu or marva or kalju (flood weir) and kalava (canal) (http://www.rainwaterharvesting.org). ***Garanda*** is a small diameter well built within a large diameter well to collect percolating water & lower-level springs.

2.2 Traditional Structures and Practices Related to Irrigation Purpose

Farmers from different regions have built various traditional structures with using basic engineering skills and unique modes, for fulfilling their irrigation and drinking needs. Table 2 shows the locally available traditional irrigation practices in different parts of India.

The following sections describe briefly the ways of using traditional water management structures in agricultural fields to tap, store, and utilize water for irrigation purposes by farmers in India.

Northern Region

Kuhls are surface channels which divert water from natural flowing streams called Khuds. It run at a higher elevation to irrigate more upstream agricultural areas than the Khud itself. A community Kuhl can irrigating an area of about 20 ha, supporting

Table 2 Traditional irrigation systems in India

Region	System	Practised in
Shortage in canal water		
North	Kuls/Kuhls	Himachal Pradesh, J&K
	Zing	Ladakh
	Arghat/waterwheels	J&K
West	Khadin	Rajasthan
	Bhandara	Maharashtra
	Phad	
East	Ahar-Pyne	Bihar
	Apatani	Arunachal Pradesh
	Cheo-ozihi	Nagaland
	Zabo/Ruza	
Central	Pat system	Madhya Pradesh
South	Anicuts	Karnataka, Tamil Nadu
Water conflicts		
East	Dongs	Assam
South	Eri system	Tamil Nadu
	Oorani	
	Keres	Karnataka
	Khazana	Goa
Water productivity		
North	Guhl	Uttarakhand
West	Naada/Bandha	Rajasthan
	Saza Kuva	
East	Bamboo drip irrigation	Meghalaya
Central	Katas/Mundas/Bandhas	Gond tribals
South	Korambus	Kerala

6–30 farmers. This system consists of a headwall of boulders across a stream to divert the water flow to the fields. With this, Kuhls and Khuds carry water to run the flour mills also, called as Gharaat (Sharma & Kanwar, 2009). ***Zings*** are the small tanks or reservoirs dug into the ground to collect glacier and snowmelt water. The system helps in conservation of water for drinking and irrigation purposes in dry summer months, recharging groundwater in Ladakh. ***Arghat/Waterwheels*** system lifted water from the river or stream via moving of a wooden or iron wheel fitted with small bucket along its rim. The collected water is then used to irrigate agricultural fields. The system is usually practiced in Jammu & Kashmir. ***Guhl*** is a standard water harvesting technique in which water is tapped from hilly slopes of Himalayas. These are 1–15 km in length and can carry a discharge of 15–100 L of water per sec (Bhattacharya, 2015).

Western Region

Khadins/Dhoras is a site-specific system which is found in Jodhpur, Barmer and Bikaner districts of Rajasthan. These can be developed in upstream drainage areas with shallow, gravelly or rocky terrain that has high runoff potential along with floodplain or gently sloping areas where the soil is suitable for crop production. The earthen embankment called Khadin bund is constructed across the gentle slope which collects the maximum possible rainwater runoff within the agricultural land. Due to water storage and chemical weathering, the Khadin soil remains moist for a long period (Bhattacharya, 2015). The ***Bhandhara*** is a water storage practice found mainly in western Deccan region. It involves constructing a permanent or temporary check dam or weir across seasonal rivers or streams to hold back flowing water. The stored water permeates into the ground, recharging wells and providing irrigation. It is a community-managed system ensuring water for drinking, farming, and livestock. ***Phad system*** is a community-managed irrigation system practiced in Dhule and Nashik districts of Maharashtra. It diverts water from rivers through Bhandharas into canals. The irrigated area is divided into blocks called Phads. ***Naada/Bandha*** is a stone check dam found in Mewar region of Rajasthan, which is built across a stream or gully to collect monsoon runoff on a stretch of land (Bhattacharya, 2015). ***Saza/Kuva*** is open pit/well having circular shape, which is constructed jointly by several farmers.

Eastern Region

Ahar-Pyne is a hydraulic structure used for managing rainwater harvesting and irrigation in areas of Chota Nagpur and Southern Bihar (Dasgupta et al., 2024). An Ahars is a rectangular shape catchment basin with three-side embankments to collect river water and rainwater from Pynes. Pynes are artificial waterways constructed to use river water in irrigating cultivable land. Multiple distributaries are created from both systems for irrigation purposes (Dasgupta et al., 2024). This system is well-suited for areas having less rainfall, rugged and rocky terrain, loose sandy soil and steep slopes. Ahar-Pyne system made paddy cultivation possible in Southern Bihar via helping farmers in water requirements (Koul et al., 2012). It also acted as a process for mitigating floods (Dasgupta et al., 2024). ***Apatani*** is an efficient water management system for rice and fish cultivation developed by Apatani tribes of Arunachal Pradesh. The system involves capturing of stream water from surrounding hills and diverted through a network of primary, secondary and tertiary channels to irrigate fields. It also harvests groundwater for irrigation. The system involves community participation for crop production (Singh & Gupta, 2002). ***Cheo*** means person responsible for laying branches and ***Ozhi*** means water. The Mezii river of Nagaland is channelized into branches and then diverted to terraces through bamboo pipes for irrigation (Srinivasan, 2022). ***Zabo/Ruza*** is a system which means impounding of water. This is practiced by Chakhesang tribes of Nagaland. Farmers store runoff water from hilltops and used for various purpose i.e. irrigation, domestic and livestock. The system consists of protected forest land on hill top, water harvesting ponds in the middle, livestock yard and the paddy field at the foot hills. ***Dongs/Dong-Bandh*** are individually owned ponds developed by the Bodo tribes of Assam to store water for irrigation which is then distributed into fields by a tool called Jahoni. In Jaintia and Khasi hills of Meghalaya, tribal farmers majorly use ***Bamboo drip irrigation***. The topography of Jaintia and Khasi is undulating and have high steep slopes with poor water holding capacity, and in this scenario the plantation of crops like arecanut and betel vine is feasible by using the water from bamboo drip irrigation. Hill top's water is directed into split bamboo pipeline or wood stands, raised 1–2 m above the ground and is distributed in orchard by installing secondary, tertiary, or additional branches of bamboo pipeline (Singh & Gupta, 2002).

Central Region

Pat system of Madhya Pradesh is a community-managed system, in which water from hilly terrain is diverted into leak-proof bunds made from mud, teak leaves and stones and then carried through water channel, which is designed to pass through deep excavated ditches and aqueducts carved into the cliffs to create a highly efficient irrigation system (Murthy et al., 2022). ***Katas/Mundas/Bandhas*** system is a traditional water harvesting system of the Gond tribe in central India (Madhya Pradesh, Orissa & Chhattisgarh). Katas is a small earthen embankment with both ends curved, constructed across slopes to capture and store rainwater for crops. Mundas are a small ponds or dug-out pits near villages/fields used for drinking water and livestock. Bandhas are a larger earthen or stone check dams constructed across streams to store water, recharge groundwater and ecological balance.

Southern Region

Eris is the system of tanks to provide enough drinking and irrigation water for the area of Tamil Nadu. ***Oorani*** are the ponds utilized by the villagers for drinking water, while the ***Kulam*** are the ponds majorly used for various temple-based activities. These were built to store rainwater and help in recharging the groundwater table, preventing soil erosion, along with wastage of runoff during monsoon, flood control systems and maintaining ecological harmony. ***Korambus*** is a structure used to increase level of water in canals along with diversion of water into fields of Kerala. These structures are built of mud, grass, coconut thatch, and brushwood. A Korambus is made by a wooden beam horizontally fixed that touches both sides of canal. It includes of several vertical wooden beams which are erected with lower ends resting firmly on the ground and their top tied to the horizontal beam. It is known as ***Chira*** in Kasargod and Thrissur districts of Kerala (http://www.rainwaterharvesting.org). ***Anicuts*** are highly significant irrigation systems of Karnataka and Tamil Nadu. It is a low, stone-based solid structure acting as a barrier to raise water level and redirect the water from a river into canals to irrigate agricultural lands. ***Keres*** are series of tanks to provide water from either Anicuts (check weir) or streams. ***Khazana*** is an irrigation system of Goa, where Zuari and Mandovi river water are used to irrigate the big rice fields and low-lying agricultural fields.

3 Legal and Policy Landscape

Water resource management in India faces significant challenges due to inadequate policies and fragmented legal frameworks. However, the National Water Policy of 2012 recognizes the critical need for decentralized, catchment-level planning, surface and groundwater conservation, and restoring/maintenance of traditional water harvesting systems as part of the Integrated Water Resources Management (Ministry of Water Resources, 2012). Pradhan Mantri Krishi Sinchayee Yojana (PMKSY) highlights the generation of new water sources and repair, restoration and revitalization of defunct water sources, and strengthening the carrying capacity of traditional village-level systems. PMKSY therefore provides programme-level entry points and finance for rehabilitating traditional harvesting for irrigation and groundwater augmentation (Ministry of Agriculture and Farmers Welfare, 2015). Jal Jeevan Mission (JJM) creating scope to integrate village ponds, recharge tanks and catchment conservation as part of long-term drinking water security plans (Ministry of Jal Shakti, 2019). Mission Amrit Sarovar emphasize both construction/renewal and long-term maintenance arrangements for institutionalizing a restoration pathway for ponds and tanks (Ministry of Rural Development, 2024). Atal Bhujal Yojana (ABY) explicitly promotes convergence with schemes that finance recharge and local water-harvesting interventions, enabling traditional structures to be part of village-level groundwater plans (Ministry of Jal Shakti, 2023). The existing water laws and policies require strengthening to address environmental, ecological, and equity issues. India's federal structure complicates water management, with inconsistencies between national and state legislation hindering effective governance (Ali, 2024). Most national schemes encourage convergence for financing restoration, but piecemeal budgeting, department-wise silos and differing sectoral mandates often limit joined-up implementation. Legal clarity on long-term ownership (community vs. department) and finance for operation and management remains a common implementation gap. Table 3 highlights the challenges in revival of traditional water harvesting systems in India.

4 Roadmap for Revival of Traditional Water Wisdom

India's path to water security necessitates a roadmap that deeply integrates traditional water wisdom with modern policy and technological advancements. While national policies like the National Water Policy, 2012 acknowledge the importance of reviving traditional water systems for climate change adaptation, there remains a need for more explicit and stringent policy frameworks, especially within environmental and disaster management policies, to effectively protect and integrate these age-old solutions. At local levels, engagement with Panchayats and leveraging the inherent community-based management of traditional systems, such as the Pani Panchayat, are critical, although addressing socio-cultural inequities that can affect

Table 3 Challenges in reviving traditional water harvesting systems in India

Challenges	Manifestation	Potential strategic response
Institutional resistance	Preference for concrete dams and canals; budgetary silos	Demonstration pilots with cost–benefit evidence; capacity building in eco-hydrology
Climate variability	Intensified drought-flood events reduce the reliability and storage efficiency	Hybrid systems with contingent storage, early-warning and groundwater buffers
Policy-technology mismatch	Top-down plans overlook local geomorphology and customary rules	Participatory GIS mapping, community water budgets, subsidiarity in decision-making
Digital documentation gaps	Erosion of traditional knowledge due to generational shift; inadequate geo-referenced data	Crowdsourced heritage mapping platforms and open-access repositories

water access is also paramount. Urban planning is showing progress, with cities like Delhi recognizing the value of traditional water bodies, but this needs to expand into comprehensive strategies across all administrative tiers to foster sustainable and equitable water management.

The revival roadmap hinges on key strategies: reforming legal and regulatory frameworks to actively support traditional practices, fostering robust community involvement by leveraging local wisdom and ownership, and seamlessly integrating these traditional methods with modern technologies like AI and advanced mapping for enhanced efficiency and resilience (Fig. 1). Crucially, this includes the preservation and rehabilitation of existing traditional infrastructure, such as tank irrigation systems and stepwells, which have historically provided sustainable water solutions. A fundamental shift in societal perception, moving from viewing water solely as a commodity to understanding it as a vital resource managed with ancient wisdom, is also essential for cultivating a collective consciousness around water conservation and ensuring a resilient 'blue future' for India.

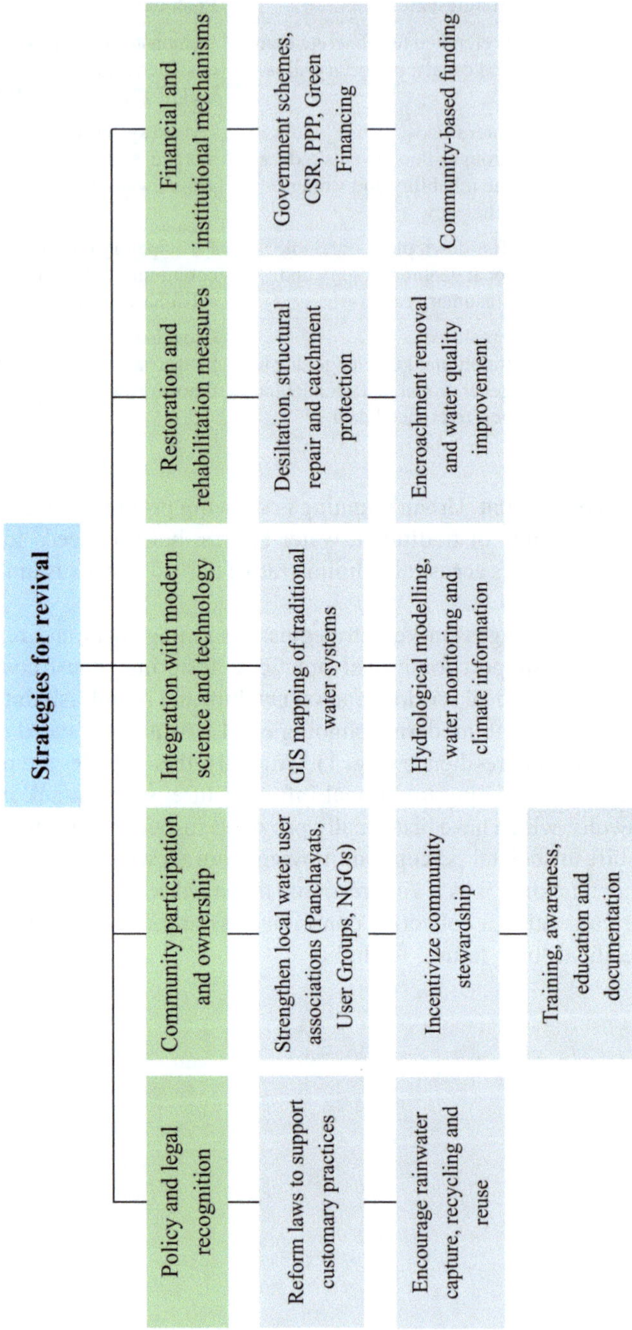

Fig. 1 Roadmap for revival of traditional water systems in India

5 Conclusion

Reviving India's traditional water wisdom is a critical pathway towards strengthening socio-ecological resilience and ensuring sustainability in response to climate change. Ecological degradation and growing water insecurity. The diverse traditional water harvesting systems from johads, ahars, tanks and stepwells to bamboo drip irrigation and zings reflect centuries of localized knowledge, community participation, and ecological harmony. However, their decline under pressures of urbanization, changing lifestyles, and policy negligence has increased both social and ecological vulnerability. Thus, by integrating these systems within the contemporary legal and policy landscape through frameworks serves as an opportunity to bridge traditional practices with modern governance. The proposed revival roadmap focused on restoration, documentation, community stewardship, integration with modern science, reforming legal and regulatory frameworks and policy incentives, offers a holistic approach that secures water resources and local economies. Ultimately, the revival of traditional water wisdom is more than a technical intervention; it is a transformative step towards fostering socio-ecological resilience, securing sustainable livelihoods, and nurturing a water-secure future rooted in both heritage and innovation.

References

Ali, I. (2024). Water legislation in India as priority aspect of water resources management. *Journal of Civil Engineering, Environment and Architecture, 71*, 47–66.

Bhandari, S., & Kaur, H. (2023). Conservation strategies for the traditional water systems: A case of Almora City, Uttarakhand. *Journal of the Institution of Engineers (India) Series A, 104*(4), 807–817. https://doi.org/10.1007/s40030-023-00759-0

Bhattacharya, S. (2015). Traditional water harvesting structures and sustainable water management in India: A socio-hydrological review. *International Letters of Natural Sciences, 37*, 30–38. https://doi.org/10.18052/www.scipress.com/ilns.37.30

Centre for Science and Environment. (2025). A look at India's water harvesting practices. Rainwater Harvesting. http://www.rainwaterharvesting.org/Rural/dec_plat_tradi.htm. Accessed September 01, 2025.

Dasgupta, R., Biswas, A., & Dutta, D. (2024). Bridging the past and present: Sustainable water management technologies through the lens of Hydro-Science in Ancient India. *Journal of Human Values*. https://doi.org/10.1177/09716858241304171

Gupta, A. K., Nair, S. S., & Singh, S. (2014). Traditional water management systems for drought mitigation in India. *Water Digest*, 36–45.

Jain, S., et al. (2024). Protecting ancient water harvesting technologies in India: Strategies for climate adaptation and sustainable development with global lessons. *Frontiers in Water, 6*, 1441365.

Koul, D. N., Singh, S., Neelam, G., & Shukla, G. (2012). Traditional water management systems—An overview of Ahar-pyne system in South Bihar plains of India and need for its revival. *Indian Journal of Traditional Knowledge*. http://nopr.niscair.res.in/bitstream/123456789/13855/1/IJTK%2011%282%29%20266-272.pdf

Ministry of Agriculture and Farmers Welfare, Government of India. (2015). Operational Guidelines of Pradhan Mantri Krishi Sinchayee Yojana (PMKSY). https://www.pmksy.gov.in/pdflinks/Guidelines_English.pdf. Accessed September 30, 2025.

Ministry of Jal Shakti, Government of India. (2019). Operational Guidelines for the Implementation of Jal Jeevan Mission. https://jaljeevanmission.gov.in/sites/default/files/guideline/JJM_Operational_Guidelines.pdf. Accessed September 30, 2025.

Ministry of Jal Shakti, Government of India. (2023). Atal Bhujal Yojana Program Guidelines Version 1.2. https://ataljal.mowr.gov.in/WriteReadData/Ataljalimages/Atal%20Bhujal%20Yojana%20Program%20Guideline%201.2.pdf. Accessed September 30, 2025.

Ministry of Rural Development, Government of India. (2024). Mission Amrit Sarovar. https://amritsarovar.gov.in/AmrutSarovarDocuments/AmritSarovarGuidelinesPhase2.pdf. Accessed September 30, 2025.

Ministry of Water Resources, Government of India. (2012). National Water Policy (2012). https://nwm.gov.in/sites/default/files/national%20water%20policy%202012_0.pdf. Accessed September 30, 2025.

Murthy, R. N. S., Srikonda, R., & Kasinath, I. V. (2022). Traditional water management systems of India. *ISVS e-Journal, 9*(2), 61–77.

Saini, J., Gupta, A. K., Dhupper, R., & Shrivastava, A. (2025). Spatio-temporal study of urban dynamics with implications on land surface temperature of Gurugram City, India. *Environmental Monitoring and Assessment, 197*(8), 1–25.

Saxena, D. (2017). Water conservation: traditional rain water harvesting systems in Rajasthan. *International Journal of Engineering Trends and Technology, 52*(2), 91–98. https://doi.org/10.14445/22315381/ijett-v52p215

Singh, P. K., Dey, P., Jain, S. K., & Mujumdar, P. (2020). Hydrology and water resources management in ancient India. *Hydrology and Earth System Sciences Discussions, 2020*, 1–20.

Singh, R. A., & Gupta, R. C. (2002). Traditional land and water management systems of North-East hill region. *Indian Journal of Traditional Knowledge, 1*(1), 32–39.

Singh, Y., Pandey, S., & Goswami, A. K. (2021). Ensuring water availability in future through revival of Indian traditional water culture. In *Environmental management—Pollution, habitat, ecology, and sustainability*. IntechOpen.

Sharma, N., & Kanwar, P. (2009). Indigenous water conservation systems—A rich tradition of rural Himachal Pradesh. *Indian Journal of Traditional Knowledge, 8*(4), 510–513. http://nopr.niscair.res.in/bitstream/123456789/6271/1/IJTK%208%284%29%20510-513.pdf

Srinivasan, G. (2022). Ancient and traditional water storage and conservation methods in north east India–A. *Editorial Board*, 19.

Sundaram, S. S., & Bhaskaran, G. (2021). A historical overview of the ancient traditional tank management system in India. *Journal of Indian History and Culture, 80.*

Yadav, S. C. (2023). Traditional rain water harvesting systems in Rajasthan: Water resources conservation and its sustainable management—A review. *Sustainability Agri, Food and Environmental Research, 11*. https://doi.org/10.7770/safer-v11n1-art2611

The HR Imperative and Smart Water Management: Linking Green Workforce Transformation with Technological Innovation

Manisha Semwal

Abstract As the shortage of water and climate change continue to worsen, new technologies are changing the way we manage and conserve water. Technology cannot fix the crisis alone, though. This chapter explores the role of human resource management (HRM) in bringing and promoting smart water solutions. It argues that a transition to a green workforce should be simultaneous with adopting smart water management. To close the gap between people and technology, this chapter employs case studies, human focus, and global best practices. It emphasizes the importance of developing a sustainable corporate culture, encouraging green leadership, and upskilling staff. HR is positioned as a strategic partner in fostering environmental change rather than as a support function.

Keywords Human resource management · Smart water management · Green workforce · Green recruitment · Employee engagement · Green leadership · Green HR practices

1 Introduction

With the worsening climate crisis, water resource management stands out as one of the most critical issues of our time. Twenty-first century issues such as urbanization, population increase, and industrial development have subjected water resources to unprecedented pressure, necessitating new innovative solutions (World Bank, 2022). Water-related issues are now soluble by virtue of technology owing to revolutionary tools in distribution, monitoring, and purification. "Smart water management" is what has been termed these advances. However, engineering solutions alone cannot be depended upon to guarantee success. As vital is human capital, particularly with regard to the development and execution of Green Human Resource Management (GHRM) policies, argue Jabbour et al. (2020).

M. Semwal (✉)
Department of Management Studies, Graphic Era (Deemed to be University), Dehradun, India

S. Pathak and M. Pandey (eds.), *Geospatial Technology and Advanced Instrumentation in Addressing Water Resources Planning and Management*, Water Informatics for Water Resource Management, https://doi.org/10.1007/978-981-92-1602-4_8

According to Renwick et al. (2013), Green HRM pertains to procedures and practices that consider the environmental effect of the company. It involves achieving corporate goals like energy efficiency and innovation while maintaining environmental stewardship. When green HRM is used in water resource management, it enables an environmentally conscious system where workforce competency and sustainability outcomes meet expectations, especially when combined with smart technology initiatives. The connection is crucial in areas with acute water scarcity, where long-term technological change and a revolutionary impact on human resources are required (Gholami et al., 2023).

The function of an HR professional is evolving as organizations increasingly focus on environmental, social, and governance (ESG) goals. They now have an important function to further green efforts in organizations beyond merely handling recruitment and training (Jackson et al., 2011). developing an innovative, conscious work culture is also a part of their responsibilities.

Take the example of smart water systems. Such technologies, based on IoT sensors and data analysis through AI, require skilled professionals capable of dealing with and interpreting complex systems. This goes on to illustrate how imperative it is that HR stimulates employees' ongoing reskilling and upskilling (Tang et al., 2018). Here, HR units are proactively driving sustainability by helping businesses adapt to and thrive in a greener and smarter future.

This chapter considers the way HR procedures can serve as the primary catalyst for the adoption of green water infrastructure and environmentally friendly technologies. Its goal is to strategically align Green Human Resource Management (GHRM) with intelligent water management programs. By highlighting the importance of capacity building, policymaking, and employee engagement, the chapter demonstrates how HR can support long-term, environmentally friendly water solutions. All things considered, this work contributes to the expanding corpus of interdisciplinary research that links technology, human resource management, and environmental science.

Digital transformation in water management is built on human capital, which includes highly qualified and creative professionals. Ignoring workforce readiness can lead to disregarded environmental goals and underuse of expensive technologies. HR directors must take the initiative to connect climate goals with human capacity building.

1.1 What is Smart Water Management

The term "smart water management" describes the application of data analytics and digital technologies to improve water quality, distribution, and conservation. To cut down on waste, guarantee fair distribution, and stop contamination, it uses automated systems, real-time monitoring, and predictive analytics. Some examples are as follows:

- Smart meters that give utilities and customers real-time feedback.
- To prevent water loss, IoT sensors in pipelines can identify leaks or changes in pressure.
- Watersheds, rainfall patterns, and possible floods can be tracked with the aid of artificial intelligence and satellite imagery.
- Machine learning algorithms to facilitate successful irrigation planning and demand forecasting.

These technologies maximize the use and conservation of water resources, lower operating costs, and improve prompt decision-making. They enable early warning for climate variability and optimize utility efficiency in both rural and urban areas.

But these technologies are more complicated than plug-and-play solutions. They require professionals with technical expertise who are also adaptable and flexible. In order to ensure that the proper people and culture are in place, HR must actively participate in managing this intricate marriage of operations, technology, and environmental goals.

1.2 The Green Workforce Revolution: An Appeal to Human Resources

To implement a new workforce management with sustainability as the theme is referred to as "green workforce transformation." It is not just a matter of generating new employment; it is also a matter of thinking differently about the old one to include skills and attitude for environmental stewardship. HR plays a key role in such cases because it decides what kind of talent is needed, organizes the employees' retraining, and links sustainability to the heart of the firm's values.

1.3 Crucial HR Procedures Include

- **Green recruitment**, which involves selecting candidates who genuinely care about sustainability in addition to possessing the necessary technical and environmental skills. It typically means working substantially with educational institutions or groups that offer green certifications and eco-focused programs.
- Providing employees the opportunity to continue their education, whether through data analysis, smart technology, or environmental regulations courses, is known as **upskilling and retraining**. This allows them to adapt to the evolving needs of sustainable work environments.
- **Leadership Construction**: Motivating mid- and senior-level executives to support sustainability. This covers guidance, exposure to international best practices, and mentoring.

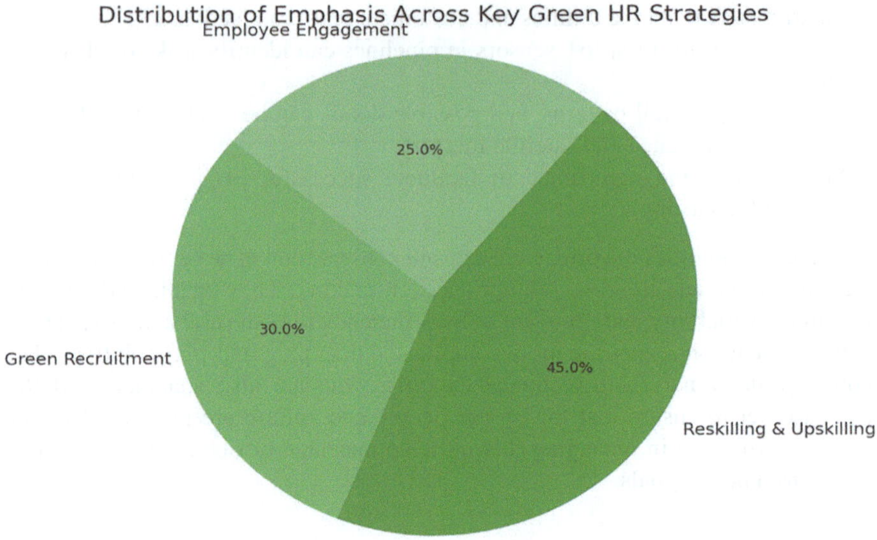

Fig. 1 Distribution of emphasis across key green HR strategies

- Developing campaigns, volunteer opportunities, and recognition programs that motivate employees to take part in water conservation and sustainability projects is known as **employee engagement**.

The following graphic displays the top three green HR strategies in in contrast highest priority sequence: green recruitment at 30%, reskilling and upskilling at 45%, and employee engagement at 25%. By concentrating on green talent management priority areas, HR teams can more efficiently distribute resources and assess impact (Fig. 1).

The strategic focus needed to create a green workforce is depicted in this image. Retraining and upskilling receive the largest allocation (45%) as organizations prioritize internalizing the transformation of current roles to meet emerging water-tech needs. The organization ensures that new hires share its values for sustainability and digital literacy by conducting 30% of its hiring process in an environmentally friendly manner. Even though employee engagement is only at 25%, it is still a crucial strategy for maintaining employees' passion and commitment to green practices.

1.4 The Function of HR in the Integration of Technology

Several disciplines must work together to implement smart water technologies, and human resources is the link that holds them all together. Implementing technology will inevitably encounter opposition, misunderstanding, or failure in the absence of a systematic approach to workforce development.

1.5 HR Can Facilitate Integration through

- **Creating Competency Models**: Outlining the abilities and conduct required for newly emerging water technology roles.
- **Multidisciplinary Research Collaboration**: Using cross-functional projects to promote communication and cooperation among engineers, IT specialists, environmental scientists, and field personnel.
- **Change Management** is the skillful and strategic handling of organizational change. Plans for communication, leadership coaching, and preparing staff to adjust to new procedures and equipment are all included in this.
- **Workforce Analytics** is the use of data to predict training gaps, identify talent needs, and assess the success of HR initiatives.

This model reveals how different HR tasks can be effectively combined with sustainability results and technology. It provides a visual framework for understanding which areas are most affected by HR (Table 1).

Professionals in human resources must act as a bridge between the needs of technology developers and the abilities of end users.

2 Boosting Smart Water Management for Humans

Technology is required to be inclusive, compassionate, and considerate of human needs. Smart water management cannot be humanized without incorporating equity, participation, and behavior change.

It entails:

- Hearing to Local Voices: Involving the community guarantees that the solutions are pertinent and considerate of cultural differences. It entails: HR is able to design frameworks for feedback, participatory planning, and local hiring.

Table 1 Distribution of emphasis across key green HR strategies

Green HR strategy	Percentage emphasis (%)	Description
Reskilling and upskilling	45	Passionate about addressing new demands in water technology by transforming current roles through focused training and development programs
Green recruitment	30	Demonstrates that the organization's commitments to sustainability and digital competency are being met in the hiring of new employees
Employee engagement	25	Concentrates on encouraging current employees' ongoing passion and dedication to ecologically friendly practices

- Developing Inclusive Solutions: Making sure women, underrepresented groups, and people with disabilities have access to education and work opportunities. Water projects that are gender inclusive have the potential to foster innovation and social justice.
- Behavioral Nudges: Utilizing behavioral science to promote conservation. HR divisions can collaborate with psychologists or sustainability specialists to create behavior-influencing campaigns, like gamified conservation challenges or incentives for conserving water.
- Implementation that is human-centered closes the gap between innovation and usability, increasing adoption rates and fostering community trust.

2.1 Case Studies: Sustainability in Water Resource Management Driven by HR

Green HR practices have been successfully incorporated into water sustainability projects by several international organizations:

- **Veolia**: Focusing on cross-functional training and reskilling for AI-enhanced water monitoring, Veolia is a world leader in water management. Smart water innovation labs and green leadership modules are part of the company's internal academy (Paulet et al., 2021). To ensure that frontline employees and technicians are proficient in using SCADA (Supervisory Control and Data Acquisition) systems, predictive analytics tools, and other smart water platforms, HR managers collaborate closely with environmental engineers to create educational materials.
- **Singapore PUB**: Through its "Watermark Champions" program, the national water agency encourages staff members to adopt water-saving practices. Its HR learning framework includes staff co-creation of sustainability projects and eco-design thinking. For example, to prototype community water-saving devices, cross-functional project teams are established, and teams that exhibit creative, user-centered results are recognized (Jabbour et al., 2020).
- **Tata Projects (India)**: Predictive analytics and automation for water management in both rural and urban areas are part of their smart water projects. HR is essential because it supports community-based hiring, facilitates multilingual toolkits, and onboards engineers through green orientation programs. From hiring to retaining employees, Tata's HR division works with local stakeholders and non-governmental organizations to instill sustainability values (Renwick et al., 2016).

The cases discussed here show how HR departments can transform smart water goals into human-driven successes rather than merely technical targets when they lead with invention, inclusivity, and deliberateness.

2.2 *Assessment and Metrics: Considering the Effects of Green HR Practices*

Maintaining green HR initiatives requires effective measurement. New KPIs that evaluate the technical, social, and environmental effects of human capital investments must be created by HR departments. Among the examples are:

- The proportion of employees with smart water instrumentation training.
- Participation of employees in innovation sprints or sustainability challenges.
- Water use reduction linked to HR-led behavior modification initiatives.
- The number of green certifications that employees possess.

Companies like Siemens and IBM have implemented HR analytics dashboards that link training attendance to lower maintenance costs, energy consumption, and water waste. Additionally, these data points can be incorporated into more comprehensive ESG (Environmental, Social, and Governance) reporting frameworks (Jabbour et al., 2020).

Furthermore, pulse surveys can be used to monitor attitudes regarding climate-related workplace initiatives, and employee engagement surveys may be altered to assess green awareness. To measure learning outcomes, HR can also conduct ecological literacy assessments both before and after training.

Businesses can show a return on investment and hasten the adoption of sustainable workforce strategies by connecting HR interventions to organizational and environmental KPIs.

3 Obstacles and the Path Ahead: Overcoming Talent and Technology Silos

Even though HR and smart water systems have the potential to work together, there are still several obstacles to overcome:

- **Lack of Technical Familiarity**: A gap with operations teams arises because many HR professionals are not familiar with concepts like telemetry, data loggers, or digital twins.
- **Change Resistance**: Veteran staff members may oppose digital disruption or doubt the benefits of training and positions with a sustainability focus.
- **Coordination Gaps**: Planning and execution are frequently disjointed because of poor communication between the engineering, IT, and HR departments.
- **Lack of Green Talent**: Hiring is difficult since the talent market for positions like circular economy strategists or water AI analysts is still developing.

HR executives need to take on roles that cross boundaries and serve as intermediaries between frontline teams, engineers, and developers in order to overcome these obstacles. Bridging knowledge gaps can also be accomplished through industry collaborations with academic institutions, attendance at international water technology expos, and joint training program development with equipment manufacturers (Renwick et al., 2013).

Going forward, the workforce landscape will be further transformed by intelligent HR technologies like blockchain-verified sustainability credentials, AI-based performance prediction tools, and digital nudges for behavioral change. Businesses will develop resilient, future-ready green workforces if they proactively incorporate these trends.

4 Conclusion

Water sustainability is not only a technical problem, but also a human one, as this chapter has shown. A convincing path to organizational and environmental resilience is provided by the combination of smart water technologies and green human resources practices. From hiring to leadership, training, and performance evaluations, HR has the resources and authority to align human capital with water stewardship. As the severity of climate change increases, the future will belong to those organizations that enable their citizens to take the lead in the green transition.

From Singapore PUB's employee-led inventions to Tata Projects' localized green hiring practices, evidence from global case studies demonstrates that HR can strategically support sustainable outcomes. By integrating green values into hiring, performance evaluation, and learning processes, HR professionals create a culture where sustainability is a practice rather than a policy.

Furthermore, decision-makers receive critical information from impact measurement metrics and frameworks, such as green HR scorecards and dashboards aligned with ESG. These tools link human behavior and resource conservation, enabling businesses to track and measure sustainability progress.

Nonetheless, HR must evolve to remain relevant, which means being proficient in digital technologies, collaborating with technology and environmental experts, and embracing a culture of continuous green innovation. The talent gap in the water industry and beyond can be closed by HR through collaborations, flexible workforce planning, and a dedication to lifelong learning.

References

Gholami, H., Rezaei, M., & Saman, M. Z. M. (2023). Role of green human resource management in environmental sustainability: A systematic review and future research direction. *Journal of Cleaner Production, 390*, 136098. https://doi.org/10.1016/j.jclepro.2023.136098

Jabbour, C. J. C., Seuring, S., de Sousa Jabbour, A. B. L., Jugend, D., & Godinho Filho, M. (2020). Unlocking the circular economy through new business models based on large-scale data: An integrative framework and research agenda. *Technological Forecasting and Social Change, 144*, 546–552. https://doi.org/10.1016/j.techfore.2019.118568

Jackson, S. E., Renwick, D. W. S., Jabbour, C. J. C., & Müller-Camen, M. (2011). State-of-the-art and future directions for green human resource management. *German Journal of Research in Human Resource Management, 25*(2), 99–116.

Paulet, E., Holland, P., & Morgan, D. (2021). A meta-analysis of employee wellbeing and green human resource management. *Human Resource Management Review, 31*(4), 100780. https://doi.org/10.1016/j.hrmr.2020.100780

Renwick, D. W. S., Jabbour, C. J. C., Muller-Camen, M., Redman, T., & Wilkinson, A. (2016). Contemporary developments in green (environmental) HRM scholarship. *The International Journal of Human Resource Management, 27*(2), 114–128. https://doi.org/10.1080/09585192.2015.1105844

Renwick, D. W. S., Redman, T., & Maguire, S. (2013). Green human resource management: A review and research agenda. *International Journal of Management Reviews, 15*(1), 1–14. https://doi.org/10.1111/j.1468-2370.2011.00328.x

Tang, G., Chen, Y., Jiang, Y., Paillé, P., & Jia, J. (2018). Green human resource management practices: Scale development and validity. *Asia Pacific Journal of Human Resources, 56*(1), 31–55. https://doi.org/10.1111/1744-7941.12147

World Bank. (2022). Water for climate resilience: A framework for action. https://openknowledge.worldbank.org

Treaty Suspension and the Water Right: Legal Limits and Strategic Necessities

Prakash Sharma, Ashish Saraswat, and Rumi Roy

Abstract The suspension of international water treaties sits at the confluence of state sovereignty, legal obligations, and human rights, particularly the water right. International law acknowledges state sovereignty and the right to address emergent difficulties while also upholding treaty obligations and human rights. This chapter examines the intricate legal framework regulating the suspension or termination of water-sharing treaties during geopolitical conflicts, environmental emergencies, or strategic adjustments. It analyses the concepts articulated in the Vienna Convention on the Law of Treaties, 1969, emphasising sections concerning significant breaches, *rebus* sic *stantibus*, and national security exceptions. Further, according to international human rights law, specifically General Comment No. 15 of the UN Committee on Economic, Social, and Cultural Rights, the suspension of water treaties may infringe upon the rights of downstream communities to water on transboundary rivers. The chapter argues that India's claim to rights over shared river waters, particularly in the context of ongoing cross-border conflicts or significant non-compliance by the other party, can be legally substantiated. India has demonstrated restraint and commitment to cooperative mechanisms under the Indus Waters Treaty, 1960 (IWT), notwithstanding continuous provocations and evolving hydrological and strategic circumstances. This chapter contends that India's assertion of national security and ecological sustainability is within the acceptable limits of international law. The chapter endorses a pragmatic and legally robust strategy that emphasises India's sovereignty and regional stability while safeguarding fundamental human rights through fair water management.

P. Sharma (✉)
Rajiv Gandhi School of Intellectual Property (RGSoIPL), IIT Kharagpur, Kharagpur, West Bengal, India
e-mail: prakash@rgsoipl.iitkgp.ac.in

A. Saraswat
National Law University, Jodhpur, Rajasthan, India

R. Roy
Vivekananda School of Law and Legal Studies (VSLLS), Vivekananda Institute of Professional Studies-Technical Campus (VIPS-TC), New Delhi, India

S. Pathak and M. Pandey (eds.), *Geospatial Technology and Advanced Instrumentation in Addressing Water Resources Planning and Management*, Water Informatics for Water Resource Management, https://doi.org/10.1007/978-981-92-1602-4_9

Keywords National security · IWT · Sovereignty · Regional stability · VCLT · Human rights

1 Introduction

The governance of transboundary water resources involves a complicated interplay between state sovereignty, international legal responsibilities, and the essential human right to water. As freshwater resources diminish and geopolitical conflicts continue, the legal issue of suspending or pausing water-sharing treaties becomes increasingly pressing. The international water-sharing agreements (Oregon State University, n.d.) exemplified by the Indus Waters Treaty, 1960 (India, Pakistan & IBRD, 1960) between India (upper riparian state) and the Islamic State of Pakistan (lower riparian state), serve as essential mechanisms for collaborative governance. The geographical and hydrological characteristics of these shared rivers are summarised in Table 1, which presents the key transboundary rivers between India and Pakistan, their origins, basin areas, and major uses. Hailed as a triumph of water diplomacy (Raghavan, 2020; Haines, 2017), IWT remained susceptible to the influences of evolving geopolitical dynamics, environmental crises, and strategic adjustments (Ranjan, 2025). The Vienna Convention on the Law of Treaties, 1969 (VCLT), provides a legal framework for the assessment of treaty suspension or termination (Vienna Convention on the Law of Treaties, 1969, Arts. 60–62). The available literature suggests that suspensions must be weighed against the commitments established by international human rights legislation. The right to water, as detailed in General Comment No. 15 by the UN Committee on Economic, Social and Cultural Rights, is deemed an essential element of the right to life and dignity (United Nations Committee on Economic, Social and Cultural Rights, 2002). The obstruction or deprivation of water access to downstream communities, whether deliberate or unintentional, poses significant issues under international human rights standards. Therefore, every suspension must be both legally valid and ethically justifiable.

This chapter meticulously analyses India's stance *vis-à-vis* the IWT (Mukul, 2025) (refer to Table 1). Despite ongoing problems from border conflicts and the alleged non-compliance, India has largely met its treaty obligations, underscoring its commitment to peaceful dispute resolution (Ranjan, 2021). It was only in the aftermath of the state-sponsored terrorist attack on civilians in Pahalgam that the Indian government announced that the IWT would be held in 'abeyance with immediate effect, until Pakistan credibly and irrevocably abjures its support for cross-border terrorism' (Ministry of External Affairs, 2025). Consequently, India reasserts its sovereign rights over transboundary river management in light of increasing geopolitical and ecological concerns. This chapter analyses the subject from a doctrinal and normative viewpoint, focusing on the legal justifications, limitations, and humanitarian implications of suspending international water treaties. It examines the tensions between a state's sovereign right to seek self-interest and its responsibilities under international law to uphold treaty obligations and protect human rights. The chapter contends

Table 1 Key transboundary rivers shared by India and Pakistan

River name	Origin	Length	Basin area(sq.km)	Downstream population dependent (approx., million)	Countries shared	Major uses	Major hazards/climate risks
Indus	Tibet (China) → India (Ladakh, J&K) → Pakistan → Arabian Sea	3180	1,165,000	237 (Indus Basin total)	China, India, Pakistan	Irrigation, hydropower (Tarbela, Mangla dams); domestic use	GLOFs (glacial Lake outburst floods); sedimentation; monsoon variability
Jhelum	India administered Kashmir	725	45,600	25	India, Pakistan	Hydropower (Mangla dam), irrigation, drinking water	Sedimentation; flash floods
Chenab	Himachal Pradesh (India)	960	67,500	40	India, Pakistan	Hydropower (Baglihar dam), irrigation canals	GLOFs; variable monsoon flows
Ravi	Himachal Pradesh (India)	720	14,442	10	India, Pakistan	Irrigation, municipal supply	Reduced baseflow from climate warming
Beas	Himachal Pradesh (India)	470	20,643	8	India	Irrigation (Bhakra dam), hydropower	Seasonal flow variability
Sutlej	Tibet (China)	1450	395,000	30	China, India, Pakistan	Hydropower (Bhakra Nangal dam), irrigation, industrial use	Sedimentation; flood peaks

Sources Data compiled from Lutz et al. (2014), Raina and Srivastava (2015), IPCC (2021), Government of India & Government of Pakistan (1960, Indus Waters Treaty). Population estimates derived from basin-level calculations (Indus Basin Initiative, 2022; World Bank, 2023)
GLOF = Glacial Lake Outburst Flood

that India's calculated yet firm position represents a legitimate exercise of its rights under international law. This approach fosters a refined and legally robust strategy that protects India's national interests, sustains regional stability, and underscores the imperative to uphold human rights, especially the right to water.

2 Conceptual Framework

The evolution of the right to water in international law demonstrates a steady transition from tacit acknowledgement to formal recognition, influenced by environmental, health, and human rights considerations. The early implications of water as a human right can be traced in Article 25 of the Universal Declaration of Human Rights, 1948 (UDHR). It mentions that '*Everyone has the right to a standard of living adequate for the health and well-being*'. A standard of living adequate for health and well-being cannot be achieved without having access to water as a human right (Meier, 2012). In 2002, the UN Committee on Economic, Social and Cultural Rights (CESCR) recognised the human right to water (Ahmad & Lilienthal, 2021). Consequently, a new interpretation of Articles 11 (right to adequate standard of living) and Article 12 (right to health) was established that includes access to sufficient, safe, accessible, and affordable water for domestic and personal use. Thus, in order to realise the right to life, health, food and housing, water is considered an essential for the realisation of other human rights.

At the same time, the notion of sovereignty and territorial integrity, widely recognised in the international system, is of utmost importance. However, states respect, protect and fulfil their obligations within their jurisdiction and refrain from taking actions that interfere with the right to water in other countries (Dang & Wang, 2025). On the other hand, VCLT's Article 60 prescribes for *'material breach'* which allows a party to suspend or terminate a treaty if the breach is in contradiction to the treaty's object and purpose. As per 'Annexure G' provision 29 appended to the IWT clearly calls for the application of other international conventions recognised by both the parties and also for the application of international customary law. The Islamic State of Pakistan's support for cross-border terrorism is directly in violation of well-established principles of the United Nations Charter, 1945—'maintenance of international peace and security'—to which both countries are signatory, and this principle is well accepted as customary international law (Kammerhofer, 2022). Therefore, it can be argued that Pakistan's promotion and support of cross-border terrorism undermines the peaceful intent of the IWT, 1966 and constitutes a material breach as per Article 60 of VCLT. This argument is further strengthened by the use of the words '*In a spirit of friendship and goodwill*' in the Preamble of IWT, 1966, which clearly stands breached in the wake of the terror attacks. Hence, this argument stands legally tenable as per the established operational principles of international law.

International law and treaty law are not stagnant in nature; they evolve with the needs of the countries and time (Gehring, 2007), and hence, there could be changes in the circumstances from the time when the treaty was signed and as per the current prevailing circumstances. This calls for taking into consideration Article 62 of VCLT, which provides for *rebus* sic *stantibus* (Fundamental Change of Circumstances), which permits parties to suspend the treaty obligations if circumstances change unpredictably and fundamentally affecting the obligations of the parties. It is justifiable to invoke this provision for putting IWT in abeyance (temporary suspension) from India's point of view, because there have been constant security threats, coupled with changing climate patterns globally, water scarcity and population growth in India. These factors can be considered as a fundamental change in circumstances when IWT was brokered, and the circumstances which are prevailing now.

3 Geopolitical and Environmental Triggers for Treaty Suspension

The management of transboundary watercourses is increasingly influenced by geopolitical instability and ecological uncertainty. Water treaties have been a common tool for cooperation in the past; however, their impact is limited when political changes and hydrological variability from climate change challenge the power of consensus on water. Despite its reputation for being resilient over time, the IWT now faces significant challenges due to unresolved compliance issues, hydrological changes caused by climate change, and increased national security demands (Zeitoun & Warner, 2006). Although the IWT was initially designed to accommodate stable hydrological conditions, it now presents a challenge in terms of its structural rigidity due to new challenges.

The suspension of the treaty by India in 2025 was initiated after the April 22nd announcement. The Pahalgam incident was not a sudden deviation but rather triggered by persistent worries about Pakistan's non-compliance under Annexure F and the inability to utilise the dispute-resolution mechanisms outlined in Article IX (Hindustan Times, 2025a, 2025b). Importantly, the suspension aimed at operational elements, such as the Permanent Indus Commission, rather than the treaty itself; this was a strategic recalibration of powers and not an absolute rejection of obligations. India's approach is consistent with Article 73 of the VCLT, which prohibits invoking essential security interests without materially breaching them. Recalibrated in academic and diplomatic terms, this 'responsible treaty realignment' combines the emerging standards of international water law, global environmental stewardship, and sustainable resource utilisation (Sinha, 2022). Unlike the previous 'Harmon Doctrine' that rejected downstream rights, India's model prioritises proportionate adjustments that are responsive to empirical hydrological data and international legal accountability (Bhandari, 2024).

Empirical studies support these hydrological points. Research on Himalayan glaciers shows that how they melt varies a lot, which has a big impact on the flow of water in the Indus Basin (Immerzeel et al., 2010). Also, predictions about sedimentation show that reservoirs are changing, which points out the limits of current agreements that do not handle these changes well (Immerzeel et al., 2010; Lutz et al., 2014; Chevuturi et al., 2022; IPCC, 2021). So, the systems in these agreements now need to be more flexible to include the real-world effects of climate change. In the bigger picture of international law, stopping or changing treaties shows how the rules are changing to balance a country's independence, human rights, and protect the environment. General Comment No. 15 from the United Nations Committee on Economic, Social and Cultural Rights (UNCESCR, 2002) clearly says that people must have access to basic water, no matter how treaties change for security reasons. So, any changes or pauses in treaties must keep water flowing enough to meet basic needs for people and follow strong legal rules that protect human rights (Sultana, 2025; Bhandari, 2024; Wouters & Chen, 2013). Looking at other international examples shows good ways to handle this. The 1994 Jordan-Israel Water Agreement uses a Joint Water Committee to adjust water sharing during dry times, keeping basic water rights and not breaking the treaty (Zeitoun & Warner, 2006; Sadoff & Grey, 2002; Dellapenna & Gupta, 2008). The Mekong Agreement from 1995 also allows for flexibility, using early warnings, tech cooperation, and adjusting water levels with the seasons to deal with water stress, without weakening the treaty. These cases show that having flexible systems within legal agreements can work well to handle changes in climate and politics.

India's decision to pause a treaty is based on careful strategic thinking, not just political moves. It shows that there are known weaknesses in the country's institutions. Government reviews found that keeping to the treaty could put water systems and energy supplies in border areas at risk. This highlights how sticking to old treaty rules can cause more problems when things change. India's action fits with international ideas like *rebus* sic *stantibus* and proportionality, which say that treaty duties can change if there are big changes in the situation. This approach needs to include care for people and the environment. New legal opinions show that keeping basic water flows between countries is important for human rights, which limits how states can use Articles 62 or 73 of the VCLT (Johnson & Van Koppen, 2023). This convergence of humanitarian, ecological and legal frameworks shapes contemporary treaty suspensions, which are increasingly justified as legally defensible when embedded within structured negotiations that prioritise ecological integrity and peremptory human rights obligations.

3.1 Geopolitical Conflicts and Strategic Necessities

Managing cross-border water resources often reflects the broader political dynamics between countries, particularly where historical tensions exist. The IWT, long regarded as a rare example of cooperation amidst conflict, stands as a testament to the possibility of sustained dialogue even in strained times rather than a relic under threat. The success of water treaties largely depends on the political and security environment, especially in regions where upstream and downstream countries have unequal power.

Emerging research on Himalayan glaciers and shifting river patterns makes it clear that water-sharing agreements like the Indus Waters Treaty must adapt to changing realities. While the IWT has withstood political turbulence, its strength will increasingly depend on how well it accounts for climate-driven shifts like faster glacial melt and altered sedimentation that are already reshaping the region's rivers. Rigid allocation rules, once effective, now need to evolve in step with the environment they were meant to govern (Mustafa, 2007). Additional data on sedimentation shifts and flow variability complicate the long-term viability of static volumetric allocations prescribed by the IWT (Paliwal, 2020). Looking at other international agreements, we can see examples of how water-sharing can change without breaking the legal structure. The 1995 Mekong Agreement allows countries like Cambodia, Laos, Thailand, and Vietnam to adjust water sharing based on real-time information and seasonal conditions. There's a joint committee that makes decisions considering both ecological and seasonal factors. This setup helps manage climate change while keeping cooperation among the countries. Similarly, the 1994 Jordan–Israel Water Treaty allows both sides to adjust water allocations every year based on reservoir levels, agricultural needs, and future water forecasts. This flexibility has helped keep collaboration going even with ongoing political disagreements (Phillips, 2012). These examples show that strong treaties are not about being rigid; instead, they are about being able to adapt to changes in climate and political situations. The Nile Basin Cooperative Framework Agreement, even though not fully accepted by all countries, shows a shift from old practices of water rights based on colonial times to more fair and environmentally friendly approaches. These agreements show a bigger trend in international law: adaptable treaty designs are not signs of instability, but rather a way to build legitimacy by incorporating rights-based approaches and taking care of the environment (Cascao, 2009). India's approach also follows these global trends. Since 2016, India has moved away from strict legal rules towards more flexible administrative actions, like managing reservoirs in steps and scheduling hydropower use. These changes helped India meet its treaty obligations while also responding effectively to challenges within the legal framework. However, this flexibility must follow human rights guidelines. The UN has stated that access to water is a basic human right that cannot be ignored. The World Health Organisation says that each person should have at least 50–100 l of water per day for health and dignity. So, any changes to water flow must ensure that downstream areas still have enough water and must not cause harm to people's well-being (Tignino, 2011). Therefore, any modulation of flows must preserve essential downstream access and be narrowly tailored to avoid humanitarian impact (Table 2).

When compared to these changing treaty structures, India's legal position shows a principled use of its own decision-making power, not a rejection of treaties. However, this power should be limited by important human rights responsibilities, especially under General Comment No. 15 of the UNCESCR (2002), which says access to water is a right that cannot be taken away. Any changes in water flow for strategic reasons

Table 2 Comparative adaptive capacity of transboundary water treaties

Treaty/Basin	Key features	Adaptive mechanisms	Climate integration	Legal basis for modification
Indus waters treaty (1960)	Bilateral treaty between India and Pakistan; rigid volumetric allocations	Lacks built-in mechanisms for revision or adaptive management	No explicit reference to climate variability	Limited; article XII allows termination only by consent
Mekong agreement (1995)	Regional agreement among the four southeast Asian countries	Includes joint committee, data-sharing, and prior notification mechanisms	Incorporates seasonal flows and ecological needs	Strong: Allows procedural evolution and technical updating
Jordan–Israel water agreement (1994)	Bilateral peace treaty with water-sharing annexe	Joint water committee enables dynamic allocation during droughts	Yes, allows annual adjustments	Moderately flexible: Adaptation via joint decision-making
Nile basin cooperative framework (draft)	Proposed multilateral agreement for the equitable use of Nile waters	Emphasises equitable and reasonable utilisation, cooperative institutions	Recognises environmental sustainability	High: Based on equitable and evolving use principles

Source Compiled by the author based on treaty texts and scholarly analyses, including Sadoff and Grey (2002), Tanzi and Arcari (2021), Brunnee and Toope (2002)

must be checked to make sure they don't harm the basic water access of people downstream. This means balancing the security exceptions in Article 73 with strong human rights rules that are above all others *(jus cogens),* to ensure that downstream communities don't lose the amount and quality of water needed for their survival and dignity (Cooley & Gleick, 2011).

India's approach to implementing water treaties must place people at the centre, safeguarding their rights to clean, reliable, and equitable access. The IWT, while a landmark in transboundary cooperation, remains silent on critical issues like water quality, seasonal flow variations, and public participation. This gap isn't just procedural; it reflects a deeper need to reimagine treaty implementation in a way that responds to lived realities and evolving environmental challenges. Without these changes, countries will keep using their own interpretations of the treaty, focusing on fairness, not causing harm, and keeping up with the main goals of the treaty. India's current stance, which is based on accepting different legal ideas, environmental needs, and protecting basic rights, shows that cooperation isn't breaking down. Instead, it's leading to a new way of handling water issues that respects rights and considers security.

3.2 Environmental Emergencies and Climate Change

Climate change has become a real and growing danger to the stability of water agreements between countries. Unlike political conflicts, which happen in bursts, the effects of climate change are ongoing and deep-rooted, slowly breaking down the water-related assumptions that these agreements are built on, such as the IWT (Hindustan Times, 2025a, 2025b). New long-term climate data show that the Indus basin is facing extreme environmental challenges: glaciers in the Himalayas are melting faster, more water is being lost through evaporation and plant use, weather patterns during the monsoon season are changing, and there's more uncertainty in water availability and weather events.

Quantitatively, glacier volume in the Hindu Kush–Himalaya area has dropped by about 30–40% since the 1970s (Maurer et al., 2019), and it is expected to lose another 36% by 2100 if emissions stay at a moderate level (Intergovernmental Panel on Climate Change, 2021). This loss is getting worse because there are more and bigger glacial lake outburst floods (GLOFs) and longer dry periods. These extreme weather events have gone up by over 60% since 1990 (Chevuturi et al., 2022). These extreme events mess up the steady flow of water, change how sediment moves, and make it harder to rely on water supplies downstream. The Indus system's water flow has become less stable, showing that old ways of dividing water volumes aren't working anymore. About 60% of the upper part of the basin's yearly water comes from snow and glaciers (Lutz et al., 2014). The delayed melting of snow and glaciers is causing problems with the timing of water needed for farming and power generation. Predictions about how much water flows during different parts of the season show that fixed water rights are no longer enough, and the idea that water will always be available year-round, as planned in mid-twentieth-century agreements, is being challenged.

The rules that control treaties about shared water between countries need to change to match the new and changing conditions of water availability. This means moving beyond the old idea of either strictly following the treaty or breaking it. One important legal idea, called *clausula rebus* sic *stantibus*, has become more important. It allows for changing or pausing treaty rules when big environmental changes affect the basic conditions that the treaty was based on (Dellapenna & Gupta, 2008). But this should be done carefully, with respect for important legal rules like *jus cogens* and the principle that no major harm should be caused. Scientists and lawyers are working more together to develop flexible ways of managing water. Rules like the Berlin Rules 2004 and the Draft Articles 2008 from the International Law Commission focus on how water flows naturally, protecting the environment, and fairness for future generations. These approaches avoid rigid methods of dividing water and instead support systems that can change with climate conditions without breaking the law (Rieu-Clarke, 2015). India's idea of adaptive sovereignty offers a balanced way between not changing treaties too much and not acting alone. It encourages regular checks on water levels, flexible ways to share water within a river basin, and plans for emergencies. This is not just a new management idea (El-Sayed et al., 2020;

Matouq et al., 2013) that can anticipate rather than merely respond to hydroclimatic disruption. It is a needed change in both science and law. Building shared abilities to track snow, melting glaciers, and water loss from plants and soil can help countries be more prepared and prevent conflicts over water.

The effects on people's well-being are huge. More than 237 million people depend on the Indus basin for their daily needs (Wouters & Chen, 2013), and any problems with getting water could cause big problems in farming, power, health, and how people make a living. International laws on human rights say that countries must not let changes in climate affect people's basic right to water. General Comment No. 15 (UNCESCR, 2002) says that having safe and enough water is a basic right that can't be taken away. The WHO says each person needs at least 50–100 l of water a day, and this is a clear measure that no reason can justify going below (United Nations Committee on Economic, Social and Cultural Rights [UNCESCR], 2002). Changes in the environment aren't just something that happens in the background; they are a major force that must be considered in how treaties are built. The IWT, as it is now, doesn't have ways to handle changes in water and climate or to adjust for big changes in the whole region or make decisions that can change as things develop. India's changing position shows a shift that is based on science, law, and morality. Climate change is no longer just a side effect; it is a key factor that affects how long treaties can last. Any good plan must bring together ecological limits, the freedom of nations, and human rights into one strong and flexible system for managing water.

4 India's Legal and Strategic Position

India embodies a nuanced equilibrium towards its international treaty commitments, developmental imperatives, and shifting geopolitical dynamics. It acknowledges her obligation to strictly abide by the letter and spirit, even when bilateral relations deteriorated. This aspect was firmly acknowledged in the 2013 Kishenganga arbitration award, which reaffirmed India's right to pursue run-of-river hydroelectric projects, provided minimum environmental flows are maintained, and strengthened India's claim of operating within treaty bounds. That said, however, in recent years, owing to persistent lack of respect and misuse of dispute mechanisms, there appears to be a growing frustration with the IWT's outdated framework. This growing dissatisfaction has prompted renewed debates on the treaty's relevance in light of changing geopolitical, environmental, and developmental realities. The following part of the chapter will briefly examine the historical background of the IWT and analyse the river basin asymmetry that underpins its structure and continuing challenges.

4.1 Historical Background and Structural Overview

India's engagement with the IWT has consistently reflected a blend of legal obligation, political pragmatism, and on-the-ground hydrological realities. Crafted in 1960 with the backing of the International Bank for Reconstruction and Development (now part of the World Bank), the treaty was envisioned not just as a water-sharing pact but as a conflict-mitigation tool in a region marked by deep-rooted tensions between India and Pakistan. However, the treaty's original legal setup, based on weather and climate data from the middle of the twentieth century and fixed borders, is now not enough to deal with the complicated changes happening today in water-related politics. These changes include melting glaciers, varying water flow from year to year, and imbalances in infrastructure across borders (Verghese, 2005). The glacier-covered areas in the Indus Basin, especially in the Western Himalayas, are the main source of the basin's continuous water supply.

Data shows that the Union Territory of Jammu and Kashmir alone provides over 60% of the water flow in the western rivers of the basin, which are the Indus, Jhelum, and Chenab. Recent studies using satellite and space-based data have found about 3136 active glaciers in the area, which makes the region highly vulnerable to changes in temperature and human activities that affect the climate (Raina & Srivastava, 2015), accentuating the systemic vulnerability to thermochroic anomalies and anthropogenic climate forces. Even though this region has a lot of water resources, India's ability to control and use this water is limited by strict rules in the treaty that set limits on how much water can be used, stored across seasons, and converted into energy (Fig. 1).

India follows strict procedures and shares water information openly as required by the Indus Waters Treaty. According to Articles VIII and IX, India has always shared information about projects before they start, exchanged water data, and allowed third-party decisions, even for projects that are disputed. Examples include the 330 MW Kishanganga and 850 MW Ratle hydroelectric projects, both of which operate within the treaty's allowed limits for non-consumptive use. These projects are built-in high-altitude areas above 2000 m and use the natural height of water to generate power, not by taking large amounts of water, which keeps them within the rules set in Article III. Even though these projects are designed to use water as it flows naturally and doesn't store much water across seasons, Pakistan raised objections that don't have strong technical support. This led to delays in the projects due to legal challenges, costing over ₹1200 crore in financial terms (Ghosh, 2013). Despite these challenges and ongoing tensions, India has kept up its communication through the Permanent Indus Commission and has not stopped working on the treaty, even when there have been political tensions.

The treaty's strict rule for changing it, Article XII (3), which requires both countries to agree, has made the whole agreement hard to change. Pakistan has used dispute rules 14 times between 1988 and 2020, mainly against Indian water projects that contribute less than 3% of the river's total yearly flow. This shows the issue is more about strategy than about water flow (Hussain & Mustafa, 2022). India's new

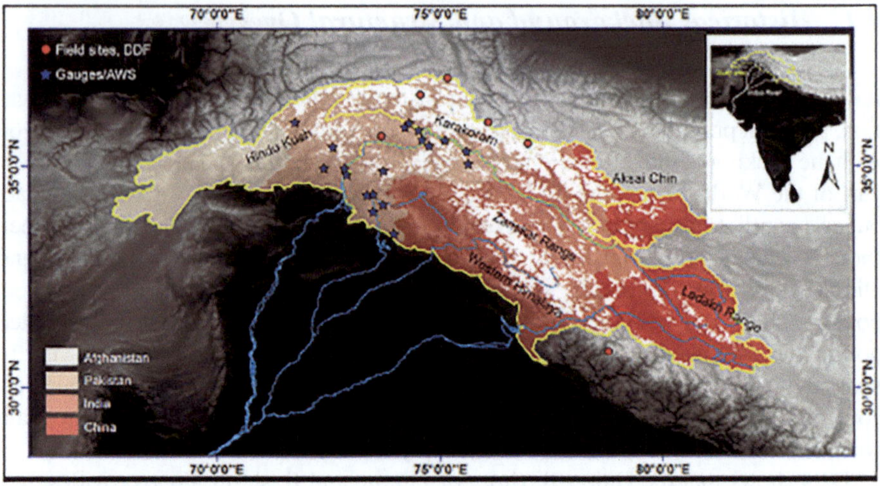

Fig. 1 Spatial distribution of glaciers across the Upper Indus Basin, highlighting key sub-basins such as Shyok, Gilgit–Hunza, Astore, and Chenab. Data support the contribution of Jammu & Kashmir to western river flows (60%). *Source* Map compiled from the Indus Waters Treaty (1960), HydroSHEDS digital elevation model (Lehner et al., 2008), and basin delineations provided by the Transboundary Freshwater Dispute Database (Oregon State University, n.d.). The cartographic base derives from the HydroSHEDS 15″ DEM dataset (Lehner et al., 2008) under a Creative Commons Attribution (CC BY 4.0) licence. The map was redrawn by the authors using ArcGIS Pro 3.2 at $\geq$ 300 dpi in CMYK format for print reproduction

approach, called Responsible Treaty Realignment, uses the ideas of necessity and proportionality from international law. This is to justify changing the treaty because of changes in glaciers and snow, and the uneven way water is used. The Western Himalayas have warmed by 0.65 °C every decade and lost 15% of their glacier mass in 40 years (Maurer et al., 2019). India says it needs to update the system based on real-time weather and water data. Instead of ending the treaty, India wants to update the way it works, making it more flexible and better suited for today's environmental challenges (Ahmed & Hassan, 2019) (Fig. 2).

4.2 *River Basin Asymmetric: Geography, Dependency, and Infrastructure*

The Indus River system exemplifies a significant disparity in geography, reliance, and infrastructural advancement between the upper and lower riparian states. India, located upstream, exerts considerable hydrological control because of its geographical advantages and sophisticated infrastructure for water storage, flood management, and hydroelectric power generation. Conversely, Pakistan's downstream position makes it significantly reliant on stable river flows for irrigation and subsistence

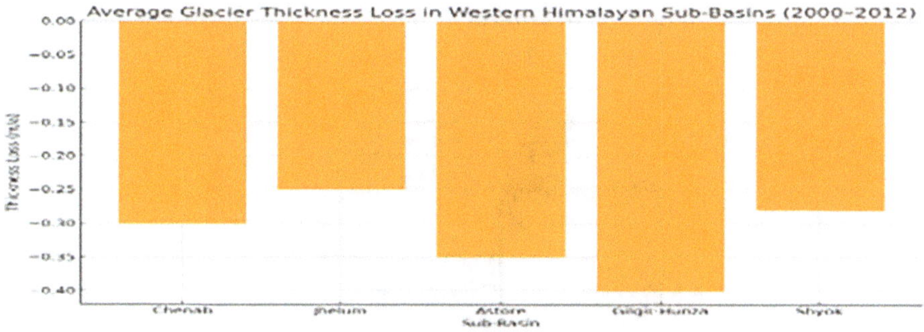

Fig. 2 Average annual glacier thickness loss (in metres per annum) from 2000 to 2012 across selected western Himalayan sub-basins of the Indus Basin. Data reflects sustained cryospheric attrition, reinforcing India's call for climate-adaptive treaty realignment. *Source* Observed trends in mean annual temperature and precipitation (1970–2020) and projected runoff change (%) under RCP 4.5 and RCP 8.5 scenarios for 2040–2060. Data derived from Raina and Srivastava (2015) and IPCC (2021). Composite visualisation prepared by the authors using open datasets; reproduced under permitted scholarly fair use and CC BY 4.0 licence. Technical specifications: 300 dpi, vector (EPS/PDF), CMYK colour profile

security, so exposing it to climate instability and infrastructural limitations. This asymmetry affects both the hydro-political interactions between the two states and the overarching patterns of water administration, allocation, and adaptation across the basin in response to changing environmental conditions.

India has acknowledged the changing environmental dynamics and stressed the need to modernise the current management framework by integrating real-time meteorological and hydrological data. Therefore, instead of pursuing the dissolution of the pact, India promotes a reassessment of its operational frameworks to improve flexibility, transparency, and resilience. This method seeks to synchronise the treaty's execution with modern difficulties presented by climate change, seasonal fluctuations, and technical progress in water monitoring and resource management (Fig. 3).

Now, the permissible use of water is defined by Article III of the IWT, which is unlimited in terms of domestic and non-consumptive usage of water; however, for agriculture, hydropower, etc. has some restrictions. This results in the negative consequences on the economy of the region (Nazakat & Nengroo, 2012), especially in the light of the agricultural dependency of the population, the hydropower sector holds a key position in meeting the energy demand in the area (Mahajan & Anthahal, 2025). As developmental projects increase, the water resource potential of the Indus River system has the potential to contribute significantly to restructuring the region's economy. Accordingly, India has invested a substantial amount of investment in water infrastructure.

When we look at the flow of Indus River Basin then it flows majorly through Pakistan i.e. 47%, then to India-39%, China—8%, and Afghanistan—6%, these are the countries that share the Indus River Basin, which spans around 1.12 million square kilometres. The Indus River begins its journey on the Tibetan Plateau close to Lake

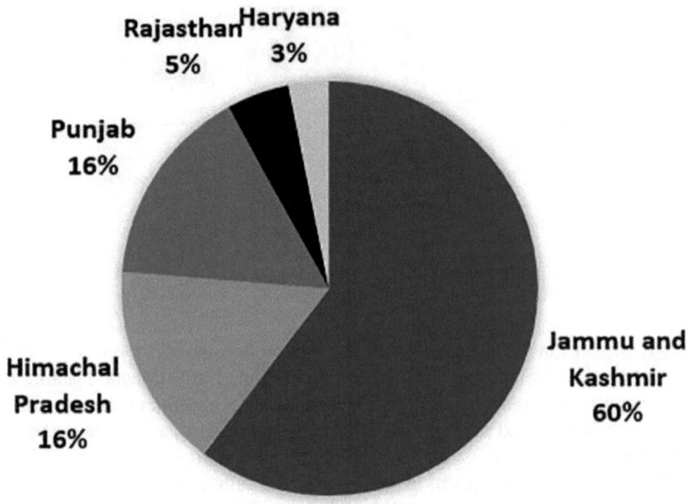

Fig. 3 Region-wise share in the Indus River Basin. *Source* Mahajan and Anthahal (2025)

Mansarovar. It then flows through the Ladakh region of India, then into Pakistan, and finally into the Arabian Sea. The Jhelum, Chenab, Ravi, Beas, and Sutlej rivers are considered to be its primary tributaries. These rivers are collectively referred to as the Eastern Rivers, and they are assigned to India. On the other hand, the Indus, Jhelum, and Chenab rivers are referred to as the Western Rivers and are allotted to Pakistan (Karki et al., 2011).

In order to regulate floods, facilitate irrigation, and generate hydroelectric power, numerous dams and barrages have been constructed along the Indus River system in Pakistan and India. Some of India's significant dams include the Bhakra Dam on the Sutlej, the Pong and Pandoh Dams on the Beas, and the Ranjit Sagar Dam also called as Thein Dam on the Ravi. Two of Pakistan's most significant dams are the Tarbela Dam on the Indus and the Mangla Dam on the Jhelum (Press Information Bureau of India, 2019). Within Pakistan, the Indus Basin is home to one of the world's largest connected irrigation systems. In order to irrigate vast agricultural lands in the Punjab and Sindh provinces, this extensive network of canals, barrages, and link canals redirects water from the Indus River and its tributaries (Van Steenbergen et al., 2015).

India has been attributed unrestricted use of the Eastern Rivers, whereas Pakistan has unrestricted access to Western Rivers as per IWT, after India's permitted use. When we look at the flow of rivers then the Western Rivers of Pakistan have a flow of approximately 135 Million Acre Feet (MAF). On the contrary, Eastern Rivers attributed to India have a flow of approximately 33 MAF (Press Information Bureau of India, 2019). Therefore, it can be validly said that Pakistan has access to 80% of water flow and India is the receiving end of only 20% of water flow, putting it in a disadvantageous position looking at its growing population and economic needs.

In particular, the areas of Jammu and Kashmir and Ladakh might be able to use more water from the Western Rivers for irrigation if India were not to uphold the Indus Water Treaty (Lad & Jaybhaye, 2023). Also, there lies a possibility to redirect the increased water resources to states in India which has altered or less rainfall, significantly boosting agricultural productivity and economic upliftment in the states such as Rajasthan, Punjab and Haryana. If IWT is suspended, then there is a scope for India to establish more hydroelectric projects on the Western Rivers. As Pakistan has most of the time raised objections to such products leading to unnecessary delays (Sharma, 2025). However, in this situation, India requires monumental diversion infrastructure, storage facilities and adequate financial flow for modifying the flow of western rivers.

From the point of view of Pakistan, the agricultural sector of Pakistan is fundamentally reliant on the waters of the Indus River system, especially the Western Rivers (Indus, Jhelum, and Chenab), which are designated to Pakistan according to the treaty. Approximately 80% of Pakistan's land used for agriculture, approximately 16 million hectares, depends on irrigation from this system (Qureshi & Perry, 2021). If India will unilaterally terminates the treaty or keep on putting the treaty in abeyance, the consequent decrease in water flow to Pakistan could possibly result in significantly diminished crop yields, extensive food shortages, and profound economic instability, especially in rural regions where agriculture underpins livelihoods.

Pakistan's constrained water storage capacity of its big dams, such as Mangla and Tarbela, which retain merely 10% of its yearly water allocation under the treaty (Rasheed & Ahmad, 2023; Ahmad et al., 2023), will even intensify its susceptibility to any interruptions in water supply in case of suspension or abeyance of IWT. Furthermore, should India choose to cease the dissemination of water flow data to Pakistan, potentially by suspending the treaty, Pakistan's capacity to strategise its irrigation policies and handle its water resources efficiently would be significantly impaired. The agricultural sector's substantial contribution to Pakistan's national income is about 20.9%, and it is the principal source of livelihood for about 65.9% of the rural population (Bhutto & Bazmi, 2007). It highlights the dependency of Pakistan on Indus River's water flow and the potential economic consequences which may arise.

Moreover, significant metropolitan centres in Pakistan, such as Karachi, Lahore, and Multan, depend primarily on the Indus River system for their water supply (Michel, 2025). A substantial reduction in water flow may result in critical drinking water shortages in many cities, perhaps inciting societal unrest. Hydropower facilities, including the Tarbela and Mangla dams, essential for Pakistan's electricity production, rely on the continuous flow of the Indus River (Sibtain et al., 2021). Decreased water flow may result in power shortages, further weakening industry and affecting the nation's overall economic stability.

4.3 India's Grounds for Treaty Reconsideration

India claimed that there has been a fundamental change of circumstances since the treaty entered into force. Although recognised under Article 62 of the VCLT, the ground is interpreted restrictively. In the *Gabcíkovo-Nagymaros* (International Court of Justice, 1997) the ICJ opined that 'the negative and conditional wording of Article 62' requires that 'the plea of fundamental change of circumstances be applied only in exceptional cases'. Considering this, India's claims in its official statement of the 'altered population demographics' and 'the need to accelerate the development of clean energy and other changes in the assumptions underlying the sharing of waters', fully constitutes for the fundamental change of circumstances (Ministry of External Affairs 2025). Further, Article 60 of the VCLT refers to the material breach, which is defined as either repudiation of the treaty or violation of a provision essential to achieving the treaty's object or purpose. It does not refer to the intensity or gravity of the breach and instead focuses on the 'character of the treaty obligation' that is being breached (Simma & Tams, 2011). The IWT, which is a technical water-sharing agreement, not only suffers legacy of disputes over the compliance with its water allocation and technical cooperation obligations, but owing to complete disregard to regional peace and support for cross-border terrorism demand call for reconsideration. Here, Article XII Clause 3 of the IWT allows for amendments, but only through mutual consent or involving World Bank mediation, if disagreement arises (Indus Waters Treaty, 1960, Art. X, cl. b). Any possibility of this appears far-fetched.

Further, states are not supposed to cause transboundary harm to another state through the use of their territory. In other words, it is the obligation of state to protect within its territory the rights of other states and is required to exercise due diligence to prevent significant harm. This means that allowing use of territory for terrorist-related activities as against India, holds the acts illegal and contrary to established principles. Nonetheless, as referred earlier, the Indus IWT, though primarily framed as a cooperative instrument for transboundary water sharing, has often found itself at the intersection of politics, security, and humanitarian considerations. While its technical provisions were designed to ensure equitable water distribution and peaceful coexistence, the treaty has also evolved into a touchstone for broader debates on justice, rights, and accountability. The invocation of the treaty during times of political hostility, particularly following terrorist incidents, reveals the tension between its apolitical design and the deeply political realities it operates within. Nonetheless, numerous individuals contend that IWT is fundamentally a technical arrangement for water distribution; however, they extend its interpretation to encompass a humanitarian framework, particularly emphasising its role in guaranteeing access to water, which is intrinsically linked to essential rights concerning life and sustenance (Singh & Krishna, 2025a, 2025b; Kebebew et al., 2025; Babu & Garimella, 2025). Yet, when India signals the suspension or restriction of treaty operations as a response to cross-border terrorism, critics argue that such actions blur the line between technical compliance and political retaliation, undermining the humanitarian essence of the agreement. Thus, the challenge lies in reconciling India's legitimate security concerns with its legal and moral commitments under an international treaty that was never intended as a political instrument but as a framework ensuring shared survival and stability in a fragile region.

4.4 Instituting a Doctrine of Responsible Treaty Realignment in Governance

Does India have to be perpetually constrained by obsolete treaty frameworks? India possesses legal options to judiciously adjust its treaty obligations. The need for responsible treaty realignment (RTR) is noteworthy here. The RTR theory advocates for a comprehensive legal and ethical framework for the renegotiation or modification of international water accords that have become outdated. It acknowledges that treaties, although based on the idea of *pacta sunt servanda*, are not unchangeable (Castle, 2024). This philosophy underscores the necessity of modifying treaty frameworks in reaction to changing hydrological, environmental, geopolitical, and technical circumstances. It promotes realignment via transparent conversation, reciprocal cooperation, and adherence to international norms, especially under the VCLT, which allows for treaty modification or suspension due to *rebus* sic *stantibus*. In other words, RTR offers a philosophical, legal, and diplomatic framework for establishing equitable, feasible, and adaptive water-sharing agreements, which fosters regional stability and sustainable resource management in the long term.

International law does not compel India to remain permanently bound by outdated treaty frameworks. Should the underlying assumptions of a treaty undergo significant alteration, such as shifts in national security concerns or the emergence of water scarcity, India may pursue revision, suspension, or withdrawal, contingent upon acting in good faith and using peaceful methods. Furthermore, the RTR concept derives its robustness from the progressive development of international water law, which increasingly prioritises fairness, sustainability, and integrated water resource management. The UN Convention on the Law of the Non-Navigational Uses of International Watercourses, 1997, despite not being ratified by all parties, underscores principles such as equitable and reasonable utilisation (United Nations, 1997, Art. 5), the duty to avoid causing significant harm (United Nations, 1997, Art. 7) and the duty to cooperate (United Nations, 1997, Art. 8). These principles, while underscoring the dynamic nature of water-sharing regimes, also align with RTR's focus on fairness and mutual benefit. Similarly, other instruments like the UN Economic Commission for Europe Convention on the Protection and Use of Transboundary Watercourses and International Lakes, 1992 (United Nations Economic Commission for Europe, 1992, Art. 9); South African Development Community Revised Protocol on Shared Watercourses, 2000 (Southern African Development Community, 2000, Art. 7); Berlin Rules on Water Resources, 2004 (International Law Association, 2004, Ch. III); Draft Legal Framework on Shared Water Resources in the

Arab Region (Draft Arab Water Convention, 2011) (Ibrahim, 2022); etc.; reflect upon the customary developments in international water law by incorporating principles of sustainability, participatory governance, and integrated water management. A comparative analysis of international and regional instruments supporting RTR is referred to in Table 3.

As water becomes a pivotal security and developmental issue due to climate change and increasing population needs, the inflexibility of outdated accords must yield to adaptable, forward-looking frameworks that address present and foreseeable difficulties. RTR advocates for mutually accepted realignments rather than unilateralism, aiming to maintain regional peace while responding to emerging realities. In this perspective, India's stance should be perceived not as a dismissal of international collaboration but as a valid assertion of its sovereign right to adjust treaty obligations in accordance with international law and ethical responsibility.

5 Need for a Balanced Framework

International law necessitates that we circumvent the notion of origins and recognise as law those laws that possess tangible effectiveness and/or political legitimacy (Franck, 1990; Hage et al., 2020). Take, for example, the ongoing 40+ years of heightened border/security tensions, which have compelled the Indian state to view water as a strategic resource. Such unilateral action falls within the accepted legal limits for the exercise of self-defence, self-help, and necessity, particularly when the armed reprisal is deemed unlawful (Reisman & Armstrong, 2006; Lowe, 2005). The action undertaken by the state is referred to as the principle of self-preservation, which is acknowledged as a customary norm within international law (Buchan, 2023). The principle permits a state to implement proportionate and non-arbitrary actions in managing strategic resources, including the transboundary water flows, provided these resources are closely related to its national security and survival interests.

Further, holding a treaty in abeyance has aligned well with India's growing strategic and developmental imperatives, which necessitate a recalibration of its engagement with IWT. India has traditionally upheld its treaty obligations with restraint and consistency; however, the recent developments indicate a doctrinal shift—a reassertion of sovereign rights to water that reflects a nuanced balance between national security, ecological sustainability, and international legal obligations. As mentioned earlier, the impact of climate change on natural resources, especially river water, in a nation that is significantly affected by climatic stress cannot be overlooked.

Table 3 Regional and International instruments supporting RTR

Instruments	Region/scope	Core principles	Mechanisms for flexibility/realignment	Relevance to RTR and India
UN Convention on the Law of the Non-Navigational Uses of International Watercourses, 1997	Global	Equitable and reasonable utilisation; no significant harm; cooperation	*Rebus* sic *stantibus*; dispute resolution; evolving obligations	Forms a normative backbone for RTR; principles can guide reassessment of treaties like the IWT
UN Economic Commission for Europe Convention on the Protection and Use of Transboundary Watercourses and International Lakes, 1992	Europe (open globally)	Prevention of transboundary impact; reasonable use; cooperation	Encourages review and updating of existing agreements; promotes joint bodies	Offers procedural models and adaptive governance mechanisms
South African Development Community Revised Protocol on Shared Watercourses (2000)	Southern Africa	Equity; no harm; sustainable use; information sharing	Explicitly allows treaty realignment to conform with protocol on shared watercourses	Strong precedent for modifying outdated water agreements under regional frameworks
Mekong Agreement, 1995	Southeast Asia	Reasonable and equitable use; sustainability; prior consultation	Institutional mechanism (Mekong River commission); provisions for adaptation	Demonstrates that treaty institutions can evolve with changing regional needs
Draft Arab Water Convention, 2011	Arab states	Sustainability; fair cooperation; periodic review	Promotes legal realignment; emphasises joint mechanisms	Shows regional appetite for updating water-sharing laws under ecological stress

(continued)

Table 3 (continued)

Instruments	Region/scope	Core principles	Mechanisms for flexibility/ realignment	Relevance to RTR and India
Berlin Rules on Water Resources, 2004	Global	Sustainability; participation; equity; prevention of harm	Encourages adaptive legal regimes; responsive governance	Reinforces RTR principles, even though non-binding; influential for judicial/academic discourse
Guadalquivir Declaration, 2005	Latin America (UNESCO-intergovernmental hydrological Programme)	Science-informed governance; shared responsibility	Promotes law-science interface; periodic legal adjustments	Reflects global trend towards legal responsiveness to environmental data

Source Compiled by the author based on the United Nations Convention on the Law of the Non-Navigational Uses of International Watercourses (1997), the UNECE Water Convention (1992), the SADC Revised Protocol on Shared Watercourses (2000), the Mekong Agreement (1995), the Draft Arab Water Convention (2011), and related scholarly analyses including Tanzi and Arcari (2021), Rieu-Clarke (2015), and Ibrahim (2022)

5.1 New Strategic Doctrine

The VCLT presently offers provisions for the suspension or termination of treaties, namely in cases of (a) material breach, (b) impossibility, and (c) *clausula rebus* sic *stantibus*. Although significant before, these tools today provide governments with only a limited legal avenue to contest or amend treaty obligations. For instance, VCLT offers no guidance on whether a treaty may be suspended or modified during conflict or heightened political tensions. Similarly, the argument of *clausula rebus* sic *stantibus* cannot be cited as grounds for terminating or suspending a treaty unless it satisfies the two-pronged test, namely: (a) the existence of those circumstances constituted an essential basis of the consent of the parties to be bound by the treaty; and (b) the effect of the change is radically to transform the extent of obligations still to be performed under the treaty. Claims that sustained cross-border terrorism constitutes an unforeseen circumstance that undermines the foundational consent to the water-treaty and would fundamentally alter the obligations will be unsubstantiated, as IWT has endured prior escalations in diplomatic relations and political turmoil (International Court of Justice, 1997, para. 104). Therefore, India's decision to place the pact in 'abeyance with immediate effect' can be regarded as 'strategic legal statecraft' (Sirohi, 2025); indicating that the treaty is not formally withdrawn but has been rendered inactive or suspended for procedural collaboration. This novel response is a legal grey zone, and draws legitimacy from customary international law doctrines of necessity (Ohlin & May, 2016) and countermeasures (International Law Commission [ILC], 2001, para. 76).

Further, such unprecedented measure of pausing the water-sharing treaty marks a departure from the previously established conventional diplomatic sanctions on trade and travel restrictions. This was against a long-standing, unaddressed wrongful state-sponsored act of terrorism that strikes at the core of trust amongst the states, which can be revived if conditions change. In this manner, India ensured that a treaty in 'abeyance' is a rare and context-specific measure. In fact, by holding the treaty in 'abeyance' and conditional (rather than permanent), India has exhibited prudence and preserved avenues for resolution. Such a stance appears to be a VCLT+ approach. This approach signifies a developing strategic doctrine in India's transboundary water diplomacy, aiming to adjust treaty obligations while adhering to them, through an ever-evolving international law.

5.2 Calibrated Model of Water Diplomacy

India's engagement with IWT reflects a resolute dedication to legal and diplomatic norms, notwithstanding bilateral challenges. India has exhibited notable caution and legal consistency in this regard. India advocated for treaty reform, emphasising the need to modernise the IWT by establishing more flexible and adaptive legal frameworks that address evolving hydrological patterns, the impending climate crisis, and future geopolitical challenges. It advocates for dynamism rather than rigidity, equity instead of entrenchment, and sustainability over political expediency. Nonetheless, this unilateral legal-diplomatic strategy was assumed blithely. Repeated occurrences of state-sponsored terrorism undermine the rationale of cooperative treaty operations (Carter & Pant, 2019).

In this context, the concept of water diplomacy, seen as essential for sustaining stable international relations, particularly in regions marked by shared resources and contested sovereignties, becomes paramount. India has unequivocally asserted that its dedication to sustaining and augmenting cooperative arrangements is contingent upon a vital condition: the cessation of cross-border terrorism. India has time and again asserted that the effectiveness of any lasting legal or diplomatic interaction, including the continued operation of the IWT, is contingent upon a basis of trust and mutual accountability, free from actions that threaten its sovereignty and internal security. The idea of reciprocity, fundamentally ingrained in international law and diplomacy, necessitates that legal collaboration is supported by a baseline of good faith.

India's current position exemplifies a measured approach to water diplomacy, wherein it upholds legal obligations, advocates for institutional reform, and promotes regional stability; however, it asserts its complete involvement on guarantees of non-aggression. India demonstrates that water accords may be both functional and flexible, rooted in legal frameworks while being attuned to security dynamics. This methodology provides a valuable framework for conflict-affected areas globally, where water diplomacy may effectively address the complex interplay of law, politics, and peace.

6 Conclusion

The cessation of international water accords, especially in unstable geopolitical regions, necessitates a delicate balance among state sovereignty, legal obligations, and humanitarian concerns. The chapter analysed India's involvement with the IWT through a sophisticated legal and normative perspective, contextualising its actions within the allowable parameters of international law. The escalating strategic demands and environmental vulnerabilities facing India require a reassessment of its transboundary water governance strategy. The claim of sovereign rights over shared water resources is based not only on the VCLT provisions, specifically regarding material breach, fundamental change of circumstances, and national security exceptions, but also on the state's broader obligation to safeguard the welfare of its populace and the integrity of its environment. In this manner, India is pioneering a VCLT+ model of treaty practice.

This position is dynamic, pragmatic, and justifiable within the context of international legal standards. It promotes a principled approach that allows the state to protect its national interests while upholding its obligations to human rights, particularly the water rights. India must consistently align its hydro-political decisions with international law while promoting regional stability through open discussions, transparency, and equitable water management. The IWT case exemplifies the essential precedent for harmonising legal flexibility, ethical responsibility, and geopolitical caution in the management of shared natural resources.

References

Ahmad, N., & Lilienthal, G. (2021). Right to water as a human right: A critical overview of international instruments. *Environmental Policy and Law, 50*(4–5), 299–308.

Ahmad, S., Jia, H., Ashraf, A., Yin, D., Chen, Z., Xu, C., Chenyang, W., Jia, Q., Xiaoyue, Z., Israr, M., & Ahmed, R. (2023). Water resources and their management in Pakistan: A critical analysis on challenges and implications. *Water-Energy Nexus, 6*, 137–150.

Ahmed, S., & Hassan, M. (2019). Institutional rigidity and strategic water management: Constraints on hydropower development under the Indus waters treaty. *International Journal of Water Resources Development, 35*(7), 1123–1142.

Babu, P., & Garimella, S. R. (2025, May 14). *India, the Indus Waters Treaty, and the limits of good faith: Abeyance, security, and the evolution of treaty practice.* Volkerrechtsblog. https://voelkerrechtsblog.org/indiathe-indus-waters-treaty-and-the-limits-of-good-faith/

Bhandari, A. (2024). Transboundary water governance and the reorientation of state practice in South Asia. *Oxford Journal of Environmental Law, 41*(2), 145–172.

Bhutto, A. W., & Bazmi, A. A. (2007). Sustainable agriculture and eradication of rural poverty in Pakistan. *Natural Resources Forum, 31*(4), 253–262.

Brunnee, J., & Toope, S. J. (2002). The changing Nile Basin regime: Does law matter? *Harvard International Law Journal, 43*(1), 105–159.

Buchan, R. (2023). Non-forcible measures and the law of self-defence. *International and Comparative Law Quarterly, 72*(1), 1–33.

Carter, D. B., & Pant, S. (2019). Terrorism and state sponsorship in world politics. In E. Chenoweth, R. English, A. Gofas, & S. K. Miller (Eds.), *The Oxford handbook of terrorism* (pp. 494–515). Oxford University Press.

Cascão, A. E. (2009). Changing power relations in the Nile River basin: Unilateralism vs. cooperation. *Water Alternatives, 2*(2), 245–268.

Castle, M. A. (2024). Renegotiating in good faith: How international treaty revisions can deepen cooperation. *The Review of International Organizations, 19*, 217–241.

Chevuturi, A., Dimri, A. P., Thayyen, R. J., & Pandit, M. K. (2022). Spatiotemporal patterns and increasing hazard frequency in the Hindu Kush Himalaya. *Environmental Research Letters, 17*(8), 084002. https://doi.org/10.1088/1748-9326/ac7ae1

Cooley, H., & Gleick, P. H. (2011). Climate-proofing transboundary water agreements. *Hydrological Sciences Journal, 56*(4), 711–718.

Dang, H., & Wang, R. Y. (2025). States' extraterritorial duty towards the right to water: Balancing sovereignty and community at the transboundary water basin level. *Journal of Energy & Natural Resources Law*, 1–22. https://www.tandfonline.com/doi/epdf/10.1080/02646811.2025.2503635?needAccess=true

Dellapenna, J. W., & Gupta, J. (2008). The evolution of global water law. In J. W. Dellapenna & J. Gupta (Eds.), *The evolution of the law and politics of water* (pp. 1–36). Springer.

El-Sayed, M. M., Hassan, M. A., & Eldeeb, A. M. (2020). Assessment of climate change impact on hydrological processes in snow-dominated river basins. *Sustainable Water Resources Management, 6*(4), 1–16.

Franck, T. M. (1990). *The power of legitimacy among nations.* Oxford University Press.

Gehring, T. (2007). Treaty-making and treaty evolution. In D. Bodansky, J. Brunnée, & E. Hey (Eds.), *The Oxford handbook of international environmental law* (pp. 467–497). Oxford University Press.

Ghosh, N. (2013). The Indus waters treaty and climate change: Monitoring and dispute resolution strategies. *Strategic Analysis, 37*(1), 21–30.

Government of India & Government of Pakistan. (1960). Indus waters treaty 1960 (with annexes). *United Nations Treaty Series, 419*, 125–290. https://treaties.un.org/doc/Publication/UNTs/Volume%20419/volume-419-I-6032-English.pdf

Hage, J., Bartels, L., & Paddeu, F. (2020). Exceptions in international law. In L. Bartels & F. Paddeu (Eds.), *Exceptions in international law* (pp. 11–34). Oxford University Press.

Haines, D. (2017). *Indus divided: India, Pakistan and the river basin dispute.* Penguin Publications.

Hindustan Times. (2025a, May 2). *Climate crisis exposes flaws in Indus pact as some rivers may dry up quicker.* Hindustan Times. https://www.hindustantimes.com/india-news/climate-crisis-exposes-flaws-in-indus-pact-as-some-rivers-may-dry-up-quicker-101746211618179.html

Hindustan Times. (2025b, June 18). *Rattled with Indus waters treaty suspension, Pakistan delegation talks 'peace' in US 'endangers livelihood'.* Hindustan Times. https://www.hindustantimes.com/world-news/rattled-with-indus-waters-treaty-suspension-pakistan-delegation-talks-peace-in-us-endangers-livelihood

Hussain, I., & Mustafa, D. (2022). Strategic invocation of dispute mechanisms in the Indus Waters Treaty: Evaluating costs and impacts. *Water International, 47*(1), 114–130.

Ibrahim, I. A. (2022). Draft legal framework for shared water resources in the Arab world: Is it really needed? *Water International, 47*(4), 512–528. https://doi.org/10.1080/02508060.2022.2076894

Immerzeel, W. W., Shrestha, A. B., & Bierkens, M. F. P. (2014). Consistent increase in high Asia's runoff due to increasing glacier melt and precipitation. *Nature Climate Change, 4*, 587–592.

Immerzeel, W. W., van Beek, L. P. H., & Bierkens, M. F. P. (2010). Climate change will affect the Asian water towers. *Science, 328*(5984), 1382–1385.

Indus Waters Treaty. (1960). *Treaty between the Government of India, the Government of Pakistan, and the International Bank for Reconstruction and Development.* https://gis.nacse.org/tfdd/tfdddocs/242ENG.pdf

Intergovernmental Panel on Climate Change. (2021). *Climate change 2021: The physical science basis. Contribution of working group I to the sixth assessment report of the IPCC.* Cambridge University Press.

International Court of Justice. (1997). Case concerning the Gabcikovo–Nagymaros project (Hungary v. Slovakia), Judgment of 25 September 1997. https://www.icj-cij.org/sites/default/files/case-related/92/092-19970925-JUD-01-00-EN.pdf

International Law Commission (ILC). (1966). Draft articles on the law of treaties with commentaries. United Nations. https://legal.un.org/ilc/texts/instruments/english/commentaries/1_1_1966.pdf

International Law Commission (ILC). (2001). Draft articles on responsibility of states for internationally wrongful acts, with commentaries. In *Yearbook of the International Law Commission*, Vol. II (Part 2) (UN Doc A/56/10; A/RES/56/83, Annex). https://legal.un.org/ilc/texts/instruments/english/commentaries/9_6_2001.pdf

Johnson, T., & van Koppen, B. (2023). Adaptive treaty mechanisms under changing hydroclimatic regimes: Transboundary freshwater cooperation in Southern Africa. *Global Environmental Politics, 23*(4), 85–107.

Kammerhofer, J. (2022). Between pragmatism and disenchantment. In P. Merkouris, J. Kammerhofer, & N. Arajärvi (Eds.), *Research handbook on the theory and practice of international lawmaking* (pp. 3–28). Cambridge University Press.

Karki, M. B., Shrestha, A. B., & Winiger, M. (2011). Enhancing knowledge management and adaptation capacity for integrated management of water resources in the Indus River Basin. *Mountain Research and Development, 31*(3), 242–251. https://doi.org/10.1659/mrd-journal-d-11-00017.1

Kebebew, T., Pellaton, C., & Tignino, M. (2025, June 16). *A treaty on the brink? India's suspension of the IWT and the case for cooperative transboundary water governance.* International Water Law Project Blog. https://www.internationalwaterlaw.org/blog/2025/06/16/a-treaty-on-the-brink-indias-suspension-of-the-iwt-andthe-case-for-cooperative-transboundary-water-governance/

Lad, R. M., & Jaybhaye, R. G. (2023, April 10). Indus water treaty: Has it prevented Jammu & Kashmir from using their own resources? *Down to Earth.* https://www.downtoearth.org.in/water/indus-water-treaty-has-it-prevented-jammu-kashmir-from-using-their-own-resources%2D%2D88689

Lehner, B., Verdin, K., & Jarvis, A. (2008). New global hydrography derived from spaceborne elevation data. *Eos, Transactions American Geophysical Union, 89*(10), 93–94.

Lowe, V. (2005). Clear and present danger: Responses to terrorism. *International and Comparative Law Quarterly, 54*(1), 185–196.

Lutz, A. F., Immerzeel, W. W., Shrestha, A. B., & Bierkens, M. F. P. (2014). Consistent increase in high Asia's runoff due to increasing glacier melt and precipitation. *Nature Climate Change, 4*, 587–592.

Mahajan, A., & Anthahal, P. C. (2025). Investment in water infrastructure and withdrawal of water in a trans-boundary river basin: A study of Jammu and Kashmir in Indus River Basin. *Current World Environment, 20*(1), 526.

Matouq, M., Al-Momani, M., Al-Quraan, S., Al-Muhtaseb, A., & Al-Dasouqi, S. (2013). The climate change implication on Jordan: A case study. *American Journal of Climate Change, 2*(1), 125–131.

Maurer, J. M., Schaefer, J. M., Rupper, S., & Yao, T. (2019). Acceleration of ice-loss across the Himalayas over the past 40 years. *Science Advances, 5*(12), eaaw0237.

Meier, B. M. (2012). *Implementing an evolving human right through water and sanitation.* https://pmc.ncbi.nlm.nih.gov/articles/PMC7006955/

Michel, D. (2025). *Can India cut off Pakistan's Indus River lifeline? Center for Strategic and International Studies (CSIS).* https://www.csis.org/analysis/can-india-cut-pakistans-indus-river-lifeline

Ministry of External Affairs. (2025, April 23). Transcript of special briefing by MEA. Government of India. https://www.mea.gov.in/media-briefings.htm?dtl/39443/Transcript_of_Special_briefing_by_MEA_April_23_2025

Mukul, S. (2025, May 30). *Not just water, money too flowed from India to Pakistan as part of Indus Treaty.* India Today. https://www.indiatoday.in/history-of-it/story/indus-waters-treaty-india-paid-pakistan-water-scarcity-damscanals-contribution-money-world-bank-deal-history-war-2733068-2025-05-30

Mustafa, D. (2007). Social construction of hydropolitics: The geographical scales of water and security in the Indus Basin. *Geographical Review, 97*(4), 484–501.

Nazakat, R., & Nengroo, A. H. (2012). Impact of Indus water treaty on J&K State: With special reference to hydro power potential. *Journal of Recent Advances in Agriculture, 1*(4).

Ohlin, J. D., & May, L. (2016). *Necessity in international law.* Oxford University Press.

Oregon State University. (n.d.). International freshwater treaties database. Transboundary freshwater dispute database, college of earth, ocean, and atmospheric sciences. Retrieved 9 October 2025, from https://transboundarywaters.ceoas.oregonstate.edu/international-freshwater-treaties-database

Paliwal, S. (2020). Indus waters treaty: Charting a new course in India–Pakistan hydro-politics. *Strategic Analysis, 44*(4), 347–361.

Phillips, D. J. H. (2012). The Jordan River basin: At the crossroads between conflict and cooperation. *International Journal of Sociology and Social Policy, 32*(1/2), 43–58.

Press Information Bureau of India. (2019). *Indus waters treaty 1960: Present status of development in India.* Ministry of Water Resources. https://www.pib.gov.in/PressReleasePage.aspx?PRID=1565906

Qureshi, A. S., & Perry, C. (2021). Managing water and salt for sustainable agriculture in the Indus Basin of Pakistan. *Sustainability, 13*(9), 5303. https://doi.org/10.3390/su13095303

Raghavan, P. (2020). *Animosity at bay: An alternative history of the India–Pakistan relationship, 1947–1952*. Oxford University Press.

Raina, V. K., & Srivastava, D. (2015). *Glaciers of the Himalaya: A shrinking heritage* (Special Publication No. 9). Geological Society of India.

Ranjan, A. (2021). *Contested waters: India's transboundary river water disputes in South Asia*. Routledge.

Ranjan, A. (2025). Indus Waters Treaty negotiations: Geopolitics, disputes and cooperation. *South Asia Research, 45*(1), 110–131.

Rasheed, M. A., & Ahmad, D. (2023). Storage and hydropower. *Global Issues in Water Policy*, 181–219. https://doi.org/10.1007/978-3-031-36131-9_7

Reisman, W. M., & Armstrong, A. (2006). The past and future of the claim of pre-emptive self-defence. *The American Journal of International Law, 100*(3), 526–553.

Rieu-Clarke, A. (2015). Transboundary hydropower projects seen through the lens of international water law. *Utrecht Law Review, 11*(1), 88–105.

Sadoff, C. W., & Grey, D. (2002). Beyond the river: The benefits of cooperation on international rivers. *Water Policy, 4*(5), 389–403.

Sharma, H. (2025, April 24). What India's suspension of the Indus waters treaty means for Pakistan and for itself. *The Indian Express*.. https://indianexpress.com/article/explained/what-indias-suspension-of-the-indus-waters-treaty-means-for-pakistan-and-for-itself-9962032/

Sibtain, M., Li, X., Bashir, H., & Azam, M. I. (2021). Hydropower exploitation for Pakistan's sustainable development: A SWOT analysis considering current situation, challenges, and prospects. *Energy Strategy Reviews, 38*, 100728.

Simma, B., & Tams, C. J. (2011). 1969 Vienna convention: Article 60- termination or suspension of the operation of a treaty as a consequence of its breach. In O. Corten & P. Klein (Eds.), *The Vienna conventions on the law of treaties: A commentary* (pp. 1351–1378). Oxford University Press.

Singh, S., & Krishna, R. S. (2025b, May 14). India, the Indus waters treaty, and the limits of good faith: Abeyance, security, and the evolution of treaty practice. *Völkerrechtsblog*. https://voelkerrechtsblog.org/india-the-indus-waters-treaty-and-the-limits-of-good-faith/

Singh, S., & Krishna, R. S. (2025a, August 26). Suspending the flow: India, the Indus treaty and the limits of international law. *Jindal Forum for International and Economic Laws*. https://jindalforinteconlaws.in/2025/08/26/suspending-the-flow-india-the-indus-treaty-and-the-limits-of-international-law/

Sinha, U. K. (2022). *Water security in India: Emerging challenges and evolving strategies*. IDSA Monograph.

Sirohi, N. (2025, April 28). Indus treaty in "abeyance": A strategic pause, not a legal breach. *Observer Research Foundation*.. https://www.orfonline.org/expert-speak/indus-treaty-in-abeyance-a-strategic-pause-not-a-legal-breach

Sultana, F. (2025). Water justice and human rights in transboundary river basins: A critical legal geography approach. *Cambridge Journal of International Law, 13*(1), 89–112.

Tanzi, A., & Arcari, M. (2021). *The United Nations convention on the Law of International Watercourses: A framework for sharing*. Brill.

Tignino, M. (2011). The right to water and sanitation in post-conflict legal mechanisms: An emerging regime? *Water International, 36*(2), 232–243.

United Nations. (1969). Vienna Convention on the Law of Treaties. *United Nations Treaty Series, 1155*, 331–679. https://treaties.un.org/doc/publication/unts/volume%201155/volume-1155-i-18232-english.pdf

United Nations. (1997). Convention on the law of the non-navigational uses of international watercourses (adopted 21 May 1997, entered into force 17 August 2014). https://treaties.un.org/doc/Treaties/1997/05/19970521%2008-42%20AM/Ch_XXVII_12.pdf.

United Nations Economic Commission for Europe. (1992). Convention on the protection and use of transboundary watercourses and international lakes (adopted 17 March 1992, entered into force 6 October 1996). Art, 9. https://unece.org/environment-policy/water/about-the-convention

United Nations Committee on Economic, Social and Cultural Rights (UNCESCR). (2002). General Comment No. 15: The right to water (Arts. 11 and 12 of the International Covenant on Economic, Social and Cultural Rights), U.N. Doc. E/C.12/2002/11. https://www.refworld.org/docid/4538838d11.html

Van Steenbergen, F., Basharat, M., & Lashari, B. K. (2015). Key challenges and opportunities for conjunctive management of surface and groundwater in mega-irrigation systems: Lower Indus, Pakistan. *Resources, 4*(4), 831–856. https://doi.org/10.3390/resources4040831

Verghese, B. G. (2005). *Waters of hope: From vision to reality in Himalaya–Ganga development cooperation.* Konark Publishers.

Wouters, P., & Chen, H. (2013). China's 'soft-path' to transboundary water cooperation examined in the light of two UN global water conventions exploring the 'Chinese way'. *Journal of Water Law, 22*(6), 229–247.

Zeitoun, M., & Warner, J. (2006). Hydro-hegemony: A framework for analysis of transboundary water conflicts. *Water Policy, 8*(5), 435–460.

Water Sustainability and Human-Tech Nexus: The Role of AI in HRM for Instrumentation-Driven Innovation

Manisha Semwal, U. K. Shukla, and Shweta

Abstract We must improve human resource management (HRM) to create a superior technology facilitator to build the water sustainability as society and technology are changing rapidly. This chapter explores how AI can impact human resource management (HRM) to address instrumental innovations, human technology relationship in sustainable water sustainability. Human capital efforts that put institutions in a position to adopt, use, and innovate with digital technologies, sensor networks, and AI-enabled solutions underpin water sustainability. This chapter considers the capabilities of HR, including talent acquisition (when recruiting employees), training and development, workforce analytics, and change management to integrate AI to help create a powerful workforce that will interface with the instrumental innovations to monitor, distribute, and preserve water. Through capital deployed in HRM processes made possible by AI and decision-making, organizations can be better equipped to respond to, and promote sustainable organizational nimbleness, and even, promote worker involvement in a culture of sustainability, all while managing the finite management of water resources sustainably. This chapter used conceptual analysis and industry expertise to provide a model of a human system-technological interface with AI to expand the idea of collaborative intelligence, or the ability of AI–human interfaces to work together to facilitate a new level of human intentionality. Overall, this chapter adds to the HRM-literature-water sustainability, in showing how creativity or flexibility and accountability can exist when one engages in one of the most valuable assets in human history.

Keywords Water sustainability · Human-Tech Nexus · Artificial intelligence · Instrumentation-driven innovation · Human resource management

M. Semwal (✉)
Department of Management Studies, Graphic Era (Deemed to be University), Dehradun, Uttarakhand, India

U. K. Shukla
CDOE Department, Koneru Lakshmaiah Education Foundation, KL (Deemed to be University), Vaddeswaram, Andhra Pradesh, India

Shweta
Chaman Lal Mahavidhalaya Autonomous Landhaura Roorkee, Haridwar, Uttarakhand, India

S. Pathak and M. Pandey (eds.), *Geospatial Technology and Advanced Instrumentation in Addressing Water Resources Planning and Management*, Water Informatics for Water Resource Management, https://doi.org/10.1007/978-981-92-1602-4_10

1 Introduction

Water, a limited resource, is invaluable in sustaining ecological balance, economic development, and human well-being. Industrialization, increased urbanization, burgeoning population, and climate change, however, are collectively threatening its sustainability at an accelerating rate. Based on the United Nations World Water Development Report (2022), more than 40% of the global population will be under severe water stress by 2050, and hence, sustainable water governance is among the most critical issues of the twenty-first century. With increasing complexity and uncertainty, conventional water management practices—relying primarily on reactive measures and infrastructure development—are no longer sufficient (Yigitcanlar et al., 2020).

At the same time, how societies observe, predict, and manage water systems is being altered by advances in digital technology, artificial intelligence (AI), Internet of Things (IoT), and big data analytics. Water management moved from reactive to predictive and adaptive approaches through AI-backed instrumentation, such as smart sensors, predictive modeling, and automation, enabling real-time decision-making, anomaly identification, and resource optimization (Krishnan et al., 2022a, 2022b). For example, cities like Singapore and Barcelona are increasingly using AI-enabled digital twins to model water networks and spot flaws before they become more serious problems (Makropoulos & Savic, 2020). These technological tools have enormous potential to improve the efficiency, equity, and climate resilience of water distribution. Organizational leadership, skill sets of employees, and the human factor doing ethical oversight of AI-based systems at the center of its integration into governance of water are critical for success. Without effective HRM practices in place, corporations can create the most technologically advanced system and transform the world of work, but they may also falsely presume that what was once adopted will not face social rejection, ethical opposition, or unwillingness to use its AI-based applications (Gelbard et al., 2023). As stated, training must be conceptualized to bridge the gap between practice, innovation, and retention. The role of hiring, change management, HR, and workforce analytics must put special thought into the alignment of a workforce to the new state-of-the-art technologies (Margherita & Braccini, 2023).

According to Joshi et al. (2025), the "Human-Tech Nexus" is a conceptual model that predicts a relationship between technological progress and human capacity to create productive outcomes. The focus is centered on collaborative intelligence, with technology working in collaboration with human judgment, not replacing it. Technology supports humans making decisions, while humans support technology by providing the context, ethical reasoning, and flexibility. This connects well with water sustainability, especially where technology allows for innovation through its tools. This is part of a flexible, people-centered ethics governance framework.

Additionally, the Human-Tech Nexus addresses a major gap in the literature on rural water management (HRM) and its application. There is nearly no literature that clarifies the relationship between the two in a unified conceptual framework. Most discussions on HRM concerning water sustainability focus on individual elements.

Much of the existing literature emphasizes technological advancements in water systems. This chapter fills this gap by discussing HRM in relation to AI-informed tools and water sustainability. More specifically, the chapter will explore how HRM can:

- Increase organizational readiness for sustainability and digital transformations.
- Encourage a culture of innovation and employee adaptability.
- Ensure that AI adoption is ethically supervised and accountable.
- Reinforce climate-sensitive water systems' resilience.

The Human-Tech Nexus provides a revolutionary approach to sustainable water governance by merging cutting-edge instrumentation systems with human resource strategies. The future is clear. Water sustainability will rely on people, organizations, and cultures once this new technology is available in the market and integrated into both social and natural systems. The challenge of water sustainability is one of the most pressing issues of the twenty-first century due to pressures from growth in industry, urbanization, population, and climate change.

To tackle these challenges, we need to examine the interactions between human technology, organizational skills, and innovative technologies. The Human-Tech Nexus provides a framework to explore how human resource management (HRM) and artificial intelligence (AI) solutions can work together to improve the resilience, agility, and innovation of water governance systems and innovation of water governance systems.

2 Managing Human Resources in the Context of Water Sustainability

The concept of water sustainability involves both technology and its impact on people or organizations. Using digital technologies, artificial intelligence (AI), and innovative tools relies on skilled workers, adaptable business models, and supportive government frameworks. This all depends on human resource management (HRM). Therefore, HRM plays a key role in connecting technological innovation with sustainable implementation (Margherita & Braccini, 2023).

3 HRM's Strategic Role in Sustainability Transitions

HRM has traditionally focused on typical administrative tasks like recruitment, performance reviews, and compliance issues. In a sustainability context, HRM takes on a strategic role. It helps promote organizational culture and supports values that aim to make sustainability part of everyday work (Jackson et al., 2020). This means that HR teams are expected to not only improve digital skills for accessing new AI technologies but also support responsible data analytics.

4 Workforce Development and Talent Acquisition

Organizations need to expand their horizons beyond just traditional engineering or environmental science expertise when it comes to hiring for water sustainability roles. Nowadays, having interdisciplinary skills that blend AI, data science, IoT, and systems thinking is becoming essential for managing modern water systems. According to a study by Upadhyay and Kumar (2022), organizations that are integrating AI into sustainability efforts require employees with hybrid skill sets—individuals who can navigate both the technical aspects and the social–ecological dynamics of these systems.

In the real world, HRM can incorporate sustainability into hiring by:

- Establishing green skills (such as circular water use and climate resilience) top priority.
- Hiring experts in digital literacy for predictive analytics and the Internet of Things.
- Attraction of multidisciplinary talent capable of working together in both technical and policy areas.

5 Change Management, AI Literacy, and Training

Employees must receive training and ongoing professional development in order to maintain their ability to adapt to changes in technology. According to studies, AI literacy—the capacity to comprehend, analyze, and interact with AI systems—is emerging as a crucial skill for a variety of sectors, including the water industry (Gelbard et al., (2023). For instance, training programs included:

- Technical modules on machine learning applications in water monitoring.
- Ethics and responsible AI adoption courses.
- Training exercises that improve flood or drought resilience planning.

The change management is also significant. Workflows are frequently disrupted by the introduction of new AI-based systems and employees who fear displacement become resistant. By addressing issues through open communication, collaborative planning, and retraining programs, HRM plays a critical role in managing organizational change (Stone & Deadrick, 2015).

6 The Essential Role of HRM in Governance, Ethics, and Accountability

Making sure that technologies are used responsibly is one of the most important concerns in the adoption of AI. HRM offers governance frameworks that support workplace justice, accountability, and openness. HR professionals are in a unique position to develop ethical standards for AI integration, according to Gelbard et al. (2023), guaranteeing adherence to privacy standards, equitable workforce distribution, and the reduction of algorithmic bias.

Stakeholder trust is also increased when ethics are incorporated into HRM. When workers feel that fairness and responsibility govern how they use digital tools, they are more inclined to adopt them (Margherita & Braccini, 2023). Therefore, HRM aims to ensure the legitimacy and acceptance of technology in addition to facilitating its adoption.

7 The Public Utilities Board (PUB) of Singapore

Singapore is renowned throughout the world for being a pioneer in sustainable and integrated water management. Thanks to its "Four National Taps" strategy and early adoption of cutting-edge technologies, the Public Utilities Board (PUB) has turned the island nation from a water-scarce country into a center of water resilience (Lee & Tan, 2021).

The success of PUB is rooted in both technology and a strong human component. Its HRM procedures show how businesses can match AI-driven tools with human capabilities:

- Strategic Workforce Planning—PUB foresees future threats like climate volatility and incorporates them into its long-term workforce planning. HRM sees to it that workers are prepared to operate with predictive models, digital twins, and real-time monitoring systems.
- Talent Development Programs—PUB spends a lot on capacity-building programs, like the Water Hub training facility, where engineers and managers are trained in both water science and digital technologies (PUB Annual Report, 2022).
- Cross-Disciplinary Teams—PUB hires not just engineers but also data scientists, IT professionals, and policy analysts, promoting interdisciplinary teams. This mechanism allows for easy incorporation of technological products into strategic choice-making.
- Ethics and Public Engagement—PUB highlights openness and public participation in water management, showing how HRM-led governance can ensure trust in technology uptake (Lee & Tan, 2021).

To ensure sustainable water management, this case demonstrates that high technological sophistication is insufficient; capacity-building HRM practices that promote adaptability and ethical stewardship are equally crucial.

8 Technological Framework: Water Sustainability Instrumentation, and Artificial Intelligence

Water sustainability is considered one of the major global challenges of the twenty-first century that needs advanced solutions based on predictive management, real-time monitoring, and intelligent decision-support capabilities based on data. The increased complexity caused by climate change, urbanization, and population stresses presents the need to replace traditional water management strategies that were largely human shaped, judgment-based, and responsive with artificial intelligence (AI), the Internet of Things (IoT), and instrumentation-based approaches (González et al., 2021). These are at the forefront of innovation based on sustainability needs in water and are changing how we track, distribute, conserve, and reuse water.

9 Smart Water Monitoring Sensors and IoT

The first step in implementing instrumentation-driven innovation is the broader use of smart sensors and the Internet of Things so that the system collects real-time data on distribution efficiency, water quality, and consumption patterns. Smart sensors are used to detect pollutants as well as gather data on pressure and leaks in water pipes; this ultimately helps water utility managers manage and reduce water losses—an important feature since distributed leakages are thought to be responsible for about 30% of global freshwater losses (World Bank, 2020).

10 Predictive Analytics and Big Data in Water Management

Large datasets, having been generated by wireless sensors and IoT devices, will have to be plumbed by very advanced analytical engines to churn out anything meaningful. Currently, machine learning tactics and big data analytics assist in anticipating supply and demand, spotting anomalies, and forming scenarios to change climate and consumption.

Predictive models can:

- Forecast droughts and floods based on hydrological and meteorological data (Sood et al., 2021).
- Maximize irrigation scheduling in agriculture, thus minimizing overconsumption (Koutroulis et al., 2023).
- Improve water allocation policies through the simulation of demand across competitive sectors (urban, industrial, agricultural).

11 Platforms for Simulation and Digital Twins

Digital twins—a virtual representation of real water systems integrating actual sensor data and an artificial-intelligence-based system representation—are one such revolutionary innovation. Digital twins allow regulators to test changes in policies or ways of operating or infrastructure without piloting the changes in the real world. (Fuller et al., 2020) For instance, water utilities can use digital twins while they do resilience planning based on modeling the effects of population, energy costs, or climate change on their water supply systems. Therefore, digital twins help with collaborative planning as well as being cost-effective.

12 Artificial Intelligence for Sustainability and Climate Resilience

The raising utilization of systems using artificial intelligence is lowering the twin risks of water shortages and extremes. With machine learning and neural networks, AI can be used to define risk watersheds, predict demand peaks, and determine variability of rainfall. As a result of climate change, such functions are especially essential in countries which are likely to suffer acute droughts or unpredictable precipitation.

Artificial intelligence systems have been applied, such as in Australia to optimize the desalination process but remain energy efficient and meet the changing needs of water (Moeini & Yazdani, 2021). Similarly, AI is used in Israel to aid drip irrigation technologies by measuring soil moisture content levels and automatically adjusting the water supply to maintain the productivity and conservation (Tal, 2020).

13 Case Study 2: The Smart Water Grid in Barcelona

Many people believe that Barcelona is leading the way in smart water management, using digital instruments to make the most of distribution and the most of sustainability. Its Smart Water Grid (SWG) unites IoT and advanced metering infrastructure (AMI) with big data platforms to monitor, as well as control water flows within the metropolitan area (González et al., 2021).

Overall, the principal characteristics of the SWG of Barcelona include:

- **Real-Time Monitoring**—Thousands of sensors monitor water quality, flow, and pressure in real time throughout the grid. Anomalies such as leakages or contamination can be detected instantly by the operators, and the response time can be reduced by days to minutes.
- **Advanced Metering Infrastructure (AMI)**—Home and business smart meters enable real-time consumption monitoring, which can be used by the customer to monitor and control consumption patterns. It has been found that AMI has reduced the residential water consumption by an average of 12% in Barcelona (Cabrera et al., 2022).
- **Predictive Maintenance**—Machine learning tools examine past trends to forecast pipe failure and schedule preventive repairs, reducing maintenance expenditures and avoiding significant water losses.
- **Citizen Participation**—The grid applications give users real-time information on consumption, leaks, and sustainability advice, encouraging change at the individual level.

The impact has been stunning: Barcelona's SWG has reduced non-revenue water losses and improved energy efficiency in pumping systems drastically, making the city a global best practice for human technology integrated water governance (González et al., 2021).

14 International Best Practices for Water Sustainability and AI

Various nations are best examples of applying AI and instrumentation-led innovation to water management:

- Israel—With worldwide recognition for its drip irrigation and desalination technologies, Israel integrates AI and IoT to maximize agricultural water consumption and wastewater recycling, with nearly 90% wastewater reuse—the highest globally (Tal, 2020).
- Australia—Artificial intelligence-enabled forecasting devices aid water allocations in the Murray–Darling Basin, harmonizing agriculture, industry, and environmental requirements under climatic uncertainty (Moeini & Yazdani, 2021).
- India's Smart Cities Mission—Urban centers such as Pune and Bhubaneswar are implementing smart water meters, IoT-based leak detection, and AI-driven demand forecasting as part of larger urban sustainability efforts (Sharma & Agarwal, 2022).
- Singapore—PUB incorporates digital twin systems for desalination plants, streamlining energy usage and promoting long-term water security (Lee & Tan, 2021).

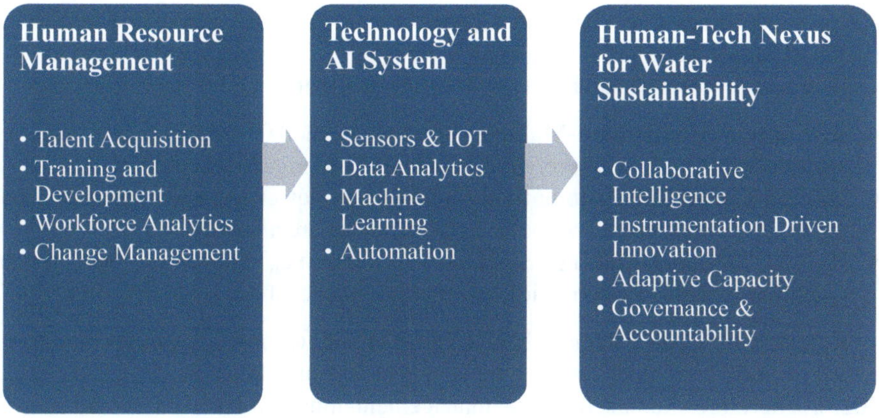

Fig. 1 Human-Tech Nexus as the convergence of HRM (human systems) and technology/AI (technical systems)

These illustrations indicate that technological innovation is not enough unless accompanied by institutional capacity, HRM policies, and citizen participation.

15 Merging Human and Technological Systems: Toward a Human-Tech Nexus Model

Although technologies like artificial intelligence (AI), IoT, and digital twins are revolutionizing water management, their performance relies on human systems that design, operate, and evolve them. In the same way, human resource management (HRM) efforts alone cannot resolve the challenges of water sustainability without technological support. The Human-Tech Nexus is intentionally marrying these two systems to create a collaborative intelligence model in which AI and instrumentation augment—not automate—the human decision-making process (Fig. 1).

It further depicts how the Human-Tech Nexus promotes water sustainability through collaborative intelligence, innovation through instrumentation, adaptive capacity, and governance.

16 Collaboration Intelligence: Enhancement, Not Replacement

The Human-Tech Nexus is based on collaborative intelligence, where human and technological intelligence complement each other. Humans provide contextual understanding, ethical decision-making, and creative/adaptive behaviors, whereas

AI provides predictive capabilities, data processing, and automated processes (Brynjolfsson & McAfee, 2017).

In regard to water sustainability, it is confirmed that:

- **AI-Supported Decision-Making**—Predictive models detect water scarcity hazards, while managers leverage contextual knowledge to schedule interventions.
- **Human-Controlled Automation**—IoT sensors detect leaks automatically, but repair crews and managers decide how resources are allocated.
- **Hybrid Decision-Making Systems**—AI identifies anomalies, but human committees analyze information and provide fair results.

Research shows that without human involvement, technology can become technocentric and out of touch with the social and environmental context (Gelbard et al., 2023). Therefore, human resource management must evolve their workforce, to manage these augmented intelligence environments.

17 The Role of HRM in Mediating AI Adoption

Human resource management is the vehicle in creating acceptance of the integration of AI and human systems. Consider:

- **Policy Formulation**—HR managers formulate ethical frameworks for AI adoption, guaranteeing privacy in data, fairness, and transparency in water management (Margherita & Braccini, 2023).
- **Skill Redesign**—Training programs advance the digital literacy required for employees to interpret AI outputs into their planning.
- **Change Management**—HR practitioners can work to reduce resistance to technological change through participatory planning and open communication (Stone & Deadrick, 2015).
- **Workforce Redesign**—AI is also changing the job roles of workers; instead of conducting manual monitoring functions, workers are switching roles toward analytic, supervisory, and innovative tasks (Upadhyay & Kumar, 2022).

18 Instrumentation, as a Connectivity Driver Innovation

Innovation driven by instrumentation uses real-time data, feedback loops, and adaptive management with IoT sensors, smart meters, and digital twins. Developing these technologies within the framework of HRM leads to cyber-physical systems where humans and machines interact seamlessly to achieve the greatest possible outcomes.

For instance:

- A sensor records increasing nitrate levels → AI diagnoses likely sources → human managers craft mitigation strategies.

- Predictive algorithm predicts drought → HRM initiates training and community participation programs to enforce adaptive policies.
- Instrumentation thus offers the technological scaffolding, while HRM offers the social scaffolding for sustainable innovation.

19 The Human-Tech Nexus Model for Water Sustainability

Drawing on the literature and practices discussed, we can think of a Human-Tech Nexus Model as being supported by four main pillars that are interlinked or connected to one another. They are as follows:

- **Collaborative Intelligence**—Synergy between humans and AI in decision-making.
- **Capacity and Skill Development**—HRM-facilitated upskilling, AI literacy, and resilience building.
- **Ethical and Governance Frameworks**—HRM integrating accountability, fairness, and transparency into technological uptake.
- **Adaptive Innovation Systems**—Feedback loops among instrumentation, analytics, and human decision-making in real time.

This framework suggests that the most effective sustainability outcomes arise when human and technological systems are designed, implemented, and evaluated in harmony with one another.

20 Governance and Ethical Oversight in the Nexus

Excessive automation may result in a variety of exclusionary, biased or unintended environmental effects; this is a significant barrier to the Human-Tech relationship. However, one option to manage those risks may be through adoption of ethical governance frameworks that emphasize human stewardship at every level of the decision-making process.

21 Practical Implications

The Human-Tech Nexus framework is a priceless conceptual tool to organizations, policymakers, and society as it proposes recommended steps and implications to not only effect the sustainability of water in the face of mounting climate change and technological resilience challenges. This framework has been formulated to shape AI implementation in the areas where it interacts with human resource management operations in a flexible, creative, and accountable manner at all levels.

21.1 Implications for the Organization: Developing Adaptive Capacity

Companies interested in advancing their HRM practices and engaging technology more actively with their human resources, cross industries. Examples of people who might seek more include utilities, agricultural cooperatives, urban planners, and regional and municipal leaders. This may involve:

- Artificial intelligence (AI) and Internet of Things (IoT) adoption led to rethink the workforce design away from simply performing routine monitoring activities, toward higher value tasks such as scenario planning, data analytics, and stakeholder engagement (Upadhyay & Kumar, 2022).
- Similarly, according to Margherita and Braccini (2023), to maintain adaptive problem-solving skills, sustainability awareness, and AI literacy of employees, organizations must put in place lifelong learning programs.
- Establishing a Culture of Innovation: Organizations can more effectively transition between simply reacting to sustainability issues and actively confronting them when they are able to establish HRM policies that encourage interdisciplinary experimentation and collaboration (Jackson et al., 2020).

This highlights that besides the installation of the most up to date technologies, it is also relevant that organizations encourage flexibility and innovativeness within human systems as a way of attaining organizational sustainability.

21.2 Relevance to HRM Systems in Practice

Effective HRM systems must be established in order to really put the Human-Tech Nexus Model into action. The following are one or two of the most important steps:

- Hiring professionals with interdisciplinary expertise in AI, water science, and sustainability management should be a top priority for HR departments looking to fill positions for hybrid skills (Upadhyay & Kumar, 2022).
- Engaging in Sustainability: HRM should integrate sustainability as a key element of organizational culture to establish collective responsibility, to which they can tie employee output indicators to water conservation initiatives (Jackson et al., 2020).
- Change Management Structures: To alleviate employee stressors about automation, effective HRM will find the balance between communication, involvement, and retraining (Stone & Deadrick, 2015).

Through these mechanisms, HRM becomes the vital element that ensures the adoption, acceptance, trust, and effective utilization of technologies.

21.3 Applications in Important Industries

There are useful applications for the Human-Tech Nexus in a variety of areas:

- Water Supply—While AI-enhanced distribution systems, digital twins, and predictive maintenance can increase operational efficiency, HRM makes sure that employees are equipped to handle these advancements.
- In agriculture, technologies like the Internet of Things (IoT) and artificial intelligence (AI) are revolutionizing precision irrigation, helping to conserve water. Meanwhile, human resource management (HRM) is empowering farmers by ensuring that they have equal access to these advancements (Tal, 2020).
- A comprehensive sustainability strategy toward urban development makes water monitoring the heart of smart cities. The HRM role plays an important role in coordinating the work of interdisciplinary teams that are building around the Water-Energy-Food (WEF) nexus (Sharma & Agarwal, 2022).

These illustrations reveal that the Human-Tech Nexus is not just a concept to be debated, but it is currently in action to promote global sustainability.

22 Conclusion

The question of the Human-Tech Nexus and the consequences of water sustainability are more relevant than ever, especially since it introduces new frames and methods. Instrumentation with AI and human resource management is created to compete with one of the most prevalent questions today. Water systems and governance are becoming more adaptive, predictive and will have monitoring and reporting requirements as they adapt further. Once again it must be noted that the potential of inventive technology would not have been augmented without a setting of human utility, corporate civilization, and human management.

This chapter aimed to demonstrate how at the heart of the Human-Tech Nexus the creation of knowledge and systems by both humans and machines is co-created and not by humans being powered by technology. Diversified AI-based water instrumentation efforts along with HRM solutions (such as the reskilling of workers, employee participation, and ethical leadership) can help companies to create transformative resilient, innovative, and sustainable water management systems. Technology as an effective tool in environmental and social governance should be intrinsically linked to HRM.

The chapter further contemplated the business, policy, and HR practitioner implications. Sustainability requires companies to assign resources to change management, workforce development, and HRM policy design. We need to establish policy-level and institutional-level governing structures to ethically regulate AI technologies in a manner that is accountable, just, and inclusive. The Human-Tech Nexus emphasizes the necessity of developing HRM systems that encourage and facilitate the development of hybrid skills and coaches a learning culture and the ethical complexity of automation.

Overall, the Human-Tech Nexus offers a meaningful format to re-examine the future of water sustainability. It proposes that the future of water governance should be more adaptive, equitable, and resilient through the integration of people and technology as in the examples of human resource management and AI support. This framework, having more clearly developed the technological innovations concerning the benefit to ecosystems and human society, also offers more means to deal with the main challenges of the climate change.

References

Brynjolfsson, E., & McAfee, A. (2017). The business of artificial intelligence: What it can—And cannot—Do for your organization. *Harvard Business Review, 95*(4), 3–11.

Cabrera, E., Pardo, M. A., & Arregui, F. (2022). Smart metering for water utilities: Lessons from Barcelona. *Urban Water Journal, 19*(5), 431–442.

Fuller, A., Fan, Z., Day, C., & Barlow, C. (2020). Digital twin: Enabling technologies, challenges, and open research. *IEEE Access, 8*, 108952–108971.

Gelbard, R., Ramon-Gonen, R., Carmeli, A., & Bittmann, R. M. (2023). Human resource management in the age of artificial intelligence: Challenges, opportunities, and a research agenda. *Journal of Business Research, 161*, 113846.

González, I., Arregui, F., & Cobacho, R. (2021). Smart water grids: Innovation in urban water distribution networks. *Water Research, 197*, 117079.

Jackson, S. E., Ones, D. S., & Dilchert, S. (2020). Managing human resources for environmental sustainability. *Human Resource Management, 59*(5), 455–470.

Joshi, S., Verma, P., & Sharma, R. (2025). Human–AI collaboration for sustainable water governance: HRM perspectives. *International Journal of Sustainable Development and Planning, 20*(4), 321–335.

Koutroulis, A. G., Papadimitriou, L. V., Grillakis, M. G., & Tsanis, I. K. (2023). Climate-smart irrigation under future water scarcity scenarios. *Agricultural Water Management, 275*, 108040.

Krishnan, R., Gupta, S., & Raj, R. (2022a). IoT and AI-enabled smart water management: A framework for sustainable resource use. *Sustainable Cities and Society, 76*, 103456.

Krishnan, S. R., Nallakaruppan, M. K., Chengoden, R., Koppu, S., Iyapparaja, M., Sadhasivam, J., & Sethuraman, S. (2022b). Smart water resource management using artificial intelligence-a review. *Sustainability, 14*(20), 13384. https://doi.org/10.3390/su142013384

Lee, T., & Tan, Y. S. (2021). Singapore's smart water management: Governance, innovation, and sustainability. *Water Policy, 23*(4), 937–951.

Margherita, A., & Braccini, A. M. (2023). Digital technologies for sustainable human resource management: A systematic literature review. *Sustainability, 15*(1), 210.

Markopoulos, C., & Savic, D. (2020). Urban water management in a changing environment: Smart technologies and sustainability. *Water, 12*(6), 1730.

Moeini, R., & Yazdani, M. (2021). Artificial intelligence in sustainable water resources management. *Environmental Modelling & Software, 145*, 105205.

PUB. (2022). *Annual Report 2021/22: Water for all—Conserve, value, enjoy*. Public Utilities Board.

Sharma, V., & Agarwal, R. (2022). Smart city water management in India: Opportunities and challenges. *Sustainable Cities and Society, 82*, 103885.

Sood, A., Bhatnagar, S., & Kumar, R. (2021). Machine learning applications for water resources: A review. *Journal of Hydrology, 598*, 126403.

Stone, D. L., & Deadrick, D. L. (2015). Challenges and opportunities affecting the future of human resource management. *Human Resource Management Review, 25*(2), 139–145.

Tal, A. (2020). Sustainability in Israel's water management: The role of innovation. *Water, 12*(6), 1767.

UNESCO World Water Assessment Programme. (2022). The United Nations world water development report 2022: Groundwater—Making the invisible visible. UNESCO. https://unesdoc.unesco.org/ark:/48223/pf0000380721

Upadhyay, A. K., & Kumar, A. (2022). Artificial intelligence and sustainable human resource management: A conceptual framework. *Journal of Cleaner Production, 357*, 131935.

World Bank. (2020). Reducing non-revenue water in developing countries: A guide for water utilities.

Yigitcanlar, T., Butler, L., & O'Connor, K. (2020). Smart technologies for sustainable water management: Urban planning and policy implications. *Journal of Environmental Planning and Management, 63*(14), 2571–2590.

MIX
Papier aus verantwortungsvollen Quellen
Paper from responsible sources
FSC® C105338

If you have any concerns about our products,
you can contact us on
ProductSafety@springernature.com

In case Publisher is established outside the EU,
the EU authorized representative is:
Springer Nature Customer Service Center GmbH
Europaplatz 3, 69115 Heidelberg, Germany

Printed by Libri Plureos GmbH
in Hamburg, Germany